T0321955

Brownian Models of Performance and Control

Brownian Models of Performance and Control covers Brownian motion and stochastic calculus at the graduate level, and illustrates the use of that theory in various application domains, emphasizing business and economics. The mathematical development is narrowly focused and briskly paced, with many concrete calculations and a minimum of abstract notation. The applications discussed include: the role of reflected Brownian motion as a storage model, queueing model, or inventory model; optimal stopping problems for Brownian motion, including the influential McDonald–Siegel investment model; optimal control of Brownian motion via barrier policies, including optimal control of Brownian storage systems; and Brownian models of dynamic inference, also called Brownian learning models, or Brownian filtering models.

J. MICHAEL HARRISON has developed and analyzed stochastic models in several different domains related to business, including mathematical finance and processing network theory. His current research is focused on dynamic models of resource sharing, and on the application of stochastic control theory in economics and operations. Professor Harrison has been honored by the Institute for Operations Research and Management Science (INFORMS) with its Expository Writing Award (1998), the Lanchester Prize for best research publication (2001), and the John von Neumann Theory Prize (2004); he was elected to the National Academy of Engineering in 2008. He is a Fellow of INFORMS and of the Institute for Mathematical Statistics.

Brownian Models of Performance and Control

J. MICHAEL HARRISON
Stanford University, California

CAMBRIDGE
UNIVERSITY PRESS

University Printing House, Cambridge CB2 8BS, United Kingdom

One Liberty Plaza, 20th Floor, New York, NY 10006, USA

477 Williamstown Road, Port Melbourne, VIC 3207, Australia

314-321, 3rd Floor, Plot 3, Splendor Forum, Jasola District Centre, New Delhi - 110025, India

79 Anson Road, #06-04/06, Singapore 079906

Cambridge University Press is part of the University of Cambridge.

It furthers the University's mission by disseminating knowledge in the pursuit of education, learning and research at the highest international levels of excellence.

www.cambridge.org
Information on this title: www.cambridge.org/9781107018396

© J. Michael Harrison 2013

First published 2013

A catalogue record for this publication is available from the British Library

ISBN 978-1-107-01839-6 Hardback

Contents

Preface

This is an expanded and updated version of a book that I published in 1985 with John Wiley and Sons, titled *Brownian Motion and Stochastic Flow Systems*. Like the original, it fits comfortably under the heading of "applied probability," its primary subjects being (i) stochastic system models based on Brownian motion, and (ii) associated methods of stochastic analysis.

Here the word "system" is used in the engineer's or economist's sense, referring to equipment, people, and operating procedures that work together with some economic purpose, broadly construed. Examples include telephone call centers, manufacturing networks, cash management operations, and data storage centers. This book emphasizes *dynamic stochastic models* of such man-made systems, that is, models in which system status evolves over time, subject to unpredictable factors like weather, demand shocks, or mechanical failures. Some of the models considered in the book are purely descriptive in nature, aimed at estimating performance characteristics like long-run average inventory or expected discounted cost, given a fixed set of system characteristics. Other models are explicitly aimed at optimizing some measure of system performance, especially through the exercise of a dynamic control capability.

Brownian models of system evolution and dynamic control are increasingly popular in economics and engineering, and in allied business fields like finance and operations. The reason for that popularity is mathematical tractability: In one instance after another, researchers working with Brownian models have been able to derive explicit solutions and clear-cut insights that were unobtainable using conventional models. In the ways that really matter, then, Brownian models provide the *simplest* possible representations of dynamic, stochastic phenomena.

Old and new content   Chapters 1, 2, 3, 4, 6, and 7 of this book correspond to chapters in the 1985 original. The last of those six "old" chapters has been substantially revised and expanded; the other five have been revised in less substantial ways. In Chapters 1 and 3 the basic properties of Brownian motion are summarized and various standard formulas are derived. Chapter 4 is devoted to the Itô calculus for Brownian motion, emphasizing Itô's formula and its various generalizations.

Chapters 2 and 6 are concerned with descriptive models of stochastic storage systems, in which an input flow and an output flow are decoupled by an intermediate storage buffer: queuing models, inventory models, and cash balance models all describe systems of that kind. Chapter 2 develops some foundational theory, and Chapter 6 is specifically concerned with storage system models based on Brownian motion, called *Brownian storage models* or *Brownian storage systems* for brevity. Formulas are developed in Chapter 6 for various standard performance measures, taking the system parameters as given.

Four dynamic control problems associated with Brownian storage models are considered in Chapter 7: optimal policies are derived under different cost structures, using both discounted and average cost optimality criteria. The subject matter of Chapter 7, like that of Chapters 2 and 6, is traditionally associated with operations research, but the monographs by Dixit (1993) and Stokey (2009) show that it is also of interest in economics.

Chapters 5, 8, and 9 are new. The first one addresses optimal stopping problems for Brownian motion, including the influential investment model of McDonald and Siegel (1986). It illustrates the guess-and-verify approach that is ubiquitous in applications, and also summarizes some insightful general theory. Chapter 8 is concerned with Brownian models of dynamic inference, in which one observes a Brownian motion whose drift rate is initially unknown, or more generally, may depend on the state of an unobserved underlying process. The problem is to make inferences about the unknown parameter or unobserved process, given the Brownian path. Learning models of this kind have a long history in statistics, where the terms "sequential analysis" and "sequential detection" are used, and also in engineering, where the central problem is described as one of "filtering" information about a parameter of interest from the Brownian noise with which it is confounded. Brownian learning models also arise increasingly in economics, especially in dynamic investment theory. Finally, Chapter 9 treats a diverse collection of examples, some involving particular applications and others more methodological in character, that further develop themes introduced in earlier chapters.

The distinctive feature of my 1985 book was its combination of compactness and concreteness: a narrow focus and brisk pace; many concrete formulas and explicit calculations; and a minimum of abstract notation. I

have made every effort to preserve that aspect of the original, and in particular, to include only such elements of general theory as are needed for the
applications considered.

Intended audience This book is intended for researchers and advanced
graduate students in economics, engineering, and operations research. As
mathematical prerequisites, readers are assumed to have knowledge of elementary real analysis, including Riemann–Stieltjes integration, at the level
of Bartle (1976), and of measure theoretic probability, including conditional expectation, at the level of Billingsley (1995). However, I have tried
to make the book accessible to readers who may lack some of the prerequisite knowledge nominally assumed. Certain essential results from probability theory and real analysis are collected in the appendices, and many
important definitions are reviewed in the text. As stated in the introduction of the 1985 original, "mathematically able readers who have at least
a nodding acquaintance with σ-algebras will be able to get by... I hope
this book will be immediately useful to readers with limited mathematical
background, and may also serve to stimulate and guide further study." The
book is aimed at non-mathematicians whose goal is to build and analyze
stochastic models.

Reflected versus regulated Brownian motion A substantial portion of
this book is devoted to a process that is called "reflected Brownian motion" in stochastic process theory. In my 1985 book I proposed the alternative name "regulated Brownian motion," arguing that the word "reflection" is confusing in this context. I used the newly coined name throughout
the 1985 book, and a number of authors, especially in economics, have
adopted my alternative terminology in their own work. However, having
won very few converts outside economics, I am reluctantly reverting to the
traditional terminology in this revision. The reason for that choice is nicely
summarized in the following statement, excerpted from an anonymous review of the book proposal that I submitted several years ago to Cambridge
University Press:

One important purpose of a textbook is to prepare the reader for the research literature,
and for better or for worse "reflected Brownian motion" is the universally accepted
standard terminology. It would be a good idea in revising the book to conform to the
standard terminology: the benefits for the reader would outweigh the aesthetic appeal
of a more imaginative but nonstandard language.

Having decided in favor of "reflected Brownian motion" for this revision, I have also adopted related terminology like "reflection mapping" and

"reflecting barrier." Despite that capitulation, the reason for my attempted revolt is worth repeating. For that purpose let $X = \{X_t, \ t \geq 0\}$ be a standard Brownian motion (zero drift and unit variance, starting at the origin) and then define

$$(0.1) \qquad\qquad Z_t := X_t - \inf_{0 \leq s \leq t} X_s, \qquad t \geq 0.$$

It has long been known that this new process Z has the same distribution as Y, where

$$(0.2) \qquad\qquad Y_t := |X_t|, \qquad t \geq 0.$$

Of course, "reflected Brownian motion" is a perfectly good name for Y, and mathematicians understandably felt that (0.2) was a more natural definition than (0.1), so Z came to be known as "an alternative representation of reflected Brownian motion." But the word "reflection" does not describe well the mapping embodied in (0.1), and it is this mapping with which one begins in applications (see Chapter 2). Moreover, we are generally interested in the situation where X is a Brownian motion with drift. Then Y and Z do *not* have the same distribution, but still Z is called "reflected Brownian motion." This terminology has even been extended to higher dimensions, where one encounters mysterious phrases like "Brownian motion with oblique reflection at the boundary." (Problem 6.13 describes a process that is usually characterized in this way.) Because the word "reflection" has the connotation of a symmetric "folding over," this terminology is potentially confusing, to say the least, but we seem to be stuck with it.

Emphasis on Itô calculus   With respect to mathematical methods, this book emphasizes Itô stochastic calculus. If one has a probabilistic model based on Brownian motion, such as the process Z defined by (0.1), then all the interesting associated quantities will be solutions of certain differential equations. For example, in this book I wish to compute expected discounted costs for various processes as functions of the starting state. To calculate such a quantity, what differential equation must be solved, and what are the appropriate boundary conditions?

Using Itô's formula, such questions can be answered systematically, which allows one to recast the original problem in purely analytic terms. Many problems can be solved by direct probabilistic means, such as the martingale methods of Chapter 3, but to solve really hard problems it is necessary to have command of both probabilistic and analytic methods.

One of my primary objectives in writing the original 1985 book was to show exactly why and how Itô's formula is so useful for solving concrete problems. Chapters 4, 6, and 7, together with their problems, have been structured with this goal in mind. I hope that even readers who have no intrinsic interest in models of buffered stochastic flow will find that the applications discussed in Chapters 6 and 7 enrich their appreciation for the general theory.

A note on organization   Readers are advised to begin with at least a quick look at the appendices. These serve not only to review prerequisite results but also to set notation and terminology. At the end of the book, just before the index, there is a list of all works cited in the text, including the pages on which they are cited. I have made no attempt to compile a comprehensive set of references on any of the subjects covered in the book, nor to suggest the relative contributions of different authors through frequency of citation.

Numbering conventions   There is a single numbering system for enunciations (lemmas, theorems, etc.) within each chapter. Thus, for example, the first three enunciations of Chapter n could be Lemma $n.1$, Definition $n.2$, and Proposition $n.3$. Equations in Chapter n are numbered $(n.1)$, $(n.2)$, etc. Similar numbering is used for enunciations and equations in the two appendices.

Acknowledgments   My initial exposure to much of the material in this book came in graduate courses from David Siegmund and Donald Iglehart, and in later interactions with David Kreps, Larry Shepp, and Rick Durrett. Erhan Çinlar read an initial draft of the 1985 original and made many helpful comments, as did Avi Mandelbaum and Ruth Williams on portions of later drafts. The original version of the book was also influenced by comments from students in courses I taught during the 1980s, including Peter Glynn, Bill Peterson, Richard Pitbladdo, Tom Sellke, and Ruth Williams.

The idea for this revised and expanded version came originally from Jan Van Mieghem, and in its preparation I have benefited from consultations with many colleagues at Stanford and elsewhere, including Barış Ata, Jim Dai, Peter DeMarzo, Darrell Duffie, Brad Efron, Brett Green, Ioannis Karatzas, Andrzej Skrzypacz, Ilya Strebulaev, Ruth Williams, and Assaf Zeevi. Among those colleagues Barış Ata deserves pride of place, having read the entire book and provided helpful suggestions on several different levels.

In the spring of 2011 I taught a graduate course at Stanford in which Joel Goh, Dmitry Orlov, Dmitry Smelov, Han-I Su, Nur Sunar, Felipe Varas Green, and Pavel Zryumov made presentations that significantly influenced the development of this book, in terms of both the new topics covered (or not covered) and the way in which those topics are treated. Finally, I am indebted to Cindy Kirby, who typed the entire book in LaTeX and expertly managed its format, and to the Stanford Graduate School of Business for financial support of this project.

Guide to Notation and Terminology

The expression $A := B$ means that A is equal to B as a matter of definition. In some sentences, the expression should be read "A, which is equal by definition to B, ..." Conditional expectations are defined only up to an equivalence. Equations involving conditional expectation, or any other random variables, should be interpreted in the almost sure sense. The terms *positive* and *increasing* are used in the weak sense, as opposed to *strictly positive* and *strictly increasing*. The equation

$$P\{X \in dx\} = f(x)\, dx$$

means that f is a density function for the random variable X. That is,

$$P\{X \in A\} = \int_A f(x)\, dx$$

for any Borel set A. In the usual way, 1_A denotes the indicator function of a set A, which equals 1 on A and equals zero elsewhere. If $(\Omega, \mathcal{F}, P)$ is a probability space and $A \in \mathcal{F}$, then 1_A is described as an indicator random variable or as the indicator of event A. To specify the time at which a stochastic process X is observed, I may write either X_t or $X(t)$ depending on the situation. On esthetic grounds, I prefer the former notation, but the latter is superior when one must write expressions like $X(T_1 + T_2)$.

Let I be an interval subset of $\mathbb{R}$ (the real line). We say that a function $f : I \to \mathbb{R}$ is C^1 (or less commonly, that f belongs to C^1) if it is continuously differentiable on the interior of I, and moreover, $f'(\cdot)$ approaches a finite limit at each closed endpoint of I, if there are any. (This state of affairs is often expressed by saying that f is continuously differentiable *up to the boundary*.) Similarly, a C^2 function on I is twice continuously differentiable on the interior of I, and its first and second derivative approach finite limits at each closed endpoint, if there are any.

Continuing in this same vein, we say that $f : I \to \mathbb{R}$ is *piecewise C^1* if it is continuously differentiable except at finitely many interior points, and moreover, $f'(\cdot)$ has finite left and right limits at each of the exceptional points, as well as a finite limit at each closed endpoint. A piecewise C^2 function is defined similarly, with both the first and second derivatives having finite left and right limits at each exceptional point, as well as finite limits at each closed endpoint. Thus, if f is piecewise C^2 on a compact interval subset of $\mathbb{R}$, both $f'(\cdot)$ and $f''(\cdot)$ are bounded.

Let f be an increasing continuous function on $[0, \infty)$. We say that f increases at a point $t > 0$ if $f(t + \epsilon) > f(t - \epsilon)$ for all $\epsilon > 0$. In this case, t is said to be a *point of increase* for f. Now let g be another continuous function on $[0, \infty)$ and consider the statement

$$f \text{ increases only when } g = 0.$$

This means that $g_t = 0$ at every point t where f increases. Many such statements appear in this book, and readers will find this terminology to be efficient if somewhat cryptic.

The last section of Appendix B discusses notational conventions for Riemann–Stieltjes integrals. As the reader will see, my general rule is to suppress the arguments of functions appearing in such integrals whenever possible. The same guiding principle is used in Chapters 4 to 6 with respect to stochastic integrals. For example, I write

$$\int_0^t X\,dW \quad \text{rather than} \quad \int_0^t X(s)\,dW(s)$$

to denote the stochastic integral of a process X with respect to a Brownian motion W. The former notation is certainly more economical, and it is also more correct mathematically, but my slavish adherence to the guiding principle may occasionally cause confusion. As an extreme example, consider the expression

$$\int_0^T e^{-\lambda t}(\Gamma - \lambda)f(Z)\,dg(X + L - U)$$

where λ is a constant, Γ is a differential operator, f and g are functions, and Z, X, L, and U are processes. This signifies the stochastic integral over $[0, T]$ of a process that has value $\exp(-\lambda t)[\Gamma f(Z_t) - \lambda f(Z_t)]$ at time t with respect to a process that has value $g(X_t + L_t - U_t)$ at time t.

The following is a list of symbols that are used with a single meaning, or at least with one dominant meaning, throughout the book. Section numbers, when given, locate either the definition of the symbol or the point of its first appearance (assuming that the appendices are read first).

$\square$	end of proof
$\wedge$ and $\vee$	maximum and minimum
$x^+ := x \vee 0$	positive part of x
$x^- := -(x \wedge 0)$	negative part of x
$\mathbb{R}$	the real line
C^1 and C^2	see text immediately above
$\mathbb{F} = \{\mathcal{F}_t,\ t \geq 0\}$	filtration (Section A.1)
$\mathcal{B}$	Borel σ-algebra on $\mathbb{R}$ (Section A.2)
$\mathcal{B}[0, \infty)$	Borel σ-algebra on $[0, \infty)$ (Section A.2)
$C := C[0, \infty)$	Section A.2
$\mathcal{C}$	Borel σ-algebra on C (Section A.2)
$\mathcal{N}(\mu, \sigma^2)$	normal distribution (Section 1.1)
$V_\beta(t)$	Wald martingale (Section 1.4)
$\Phi(x)$	$\mathcal{N}(0, 1)$ distribution function (Section 1.7)
P_x and E_x	Section 3.1
$\Gamma f := \mu f' + \frac{1}{2}\sigma^2 f''$	Section 3.3
$\alpha_1(\lambda)$ and $\alpha_2(\lambda)$	Section 3.3
$\psi_1(x)$ and $\psi_2(x)$	Section 3.3
$\theta_1(x)$ and $\theta_2(x)$	Section 3.3
$E(X; A)$	partial expectation (Section 3.3)
H	Section 4.1
$I_t(X)$	stochastic integral (Section 4.1)
H^2	Section 4.2
L^2	Section 4.2
S^2	Section 4.2
RCLL	right-continuous with left limits (Section 4.8)

1

Brownian Motion

The initial sections of this chapter are devoted to the definition of Brownian motion (the mathematical object, not the physical phenomenon) and a compilation of its basic properties. The properties in question are quite deep, and readers will be referred elsewhere for proofs. Later sections are devoted to the derivation of further properties and to calculation of several interesting distributions associated with Brownian motion.

Before proceeding, readers are advised to at least look through Appendices A and B, which enunciate some standing assumptions (in particular, joint measurability and right-continuity of stochastic processes) and explain several important conventions regarding notation and terminology. As noted there, and in the Guide to Notation and Terminology, the value of a stochastic process X at time t may be written either as X_t or as $X(t)$, depending on convenience. The former notation is generally preferred, but the latter is used when necessary to avoid clumsy typography like subscripts on subscripts.

1.1 Wiener's theorem

A stochastic process X is said to have *independent increments* if the random variables $X(t_0), X(t_1) - X(t_0), \ldots, X(t_n) - X(t_{n-1})$ are independent for any $n \geq 1$ and $0 \leq t_0 < \cdots < t_n < \infty$. It is said to have *stationary* independent increments if moreover the distribution of $X(t) - X(s)$ depends only on $t - s$. Finally, we write $Z \sim \mathcal{N}(\mu, \sigma^2)$ to mean that the random variable Z has the normal distribution with mean μ and variance σ^2. A *standard Brownian motion*, or *Wiener process*, is then defined as a stochastic process X having continuous sample paths, stationary independent increments, and $X(t) \sim \mathcal{N}(0, t)$. Thus, in our terminology, a standard Brownian motion starts at level zero almost surely. A process Y will be called a (μ, σ) Brownian motion if it has the form $Y(t) = Y(0) + \mu t + \sigma X(t)$, where X is a Wiener process and $Y(0)$ is independent of X. It follows that $Y(t + s) - Y(t) \sim$

$\mathcal{N}(\mu s, \sigma^2 s)$. We call μ and σ^2 the *drift* and *variance* of Y, respectively. The term *Brownian motion*, without modifier, will be used to embrace all such processes Y.

There remains the question of whether standard Brownian motion exists and whether it is in any sense unique. That is the subject of Wiener's theorem. For its statement, let $\mathcal{C}$ be the Borel σ-algebra on $C := C[0, \infty)$ as in Section A.2, and let X be the coordinate process on C as in Section A.3. The following is proved in the setting of $C[0, 1]$ in Section 8 of Billingsley (1999); the extension to $C[0, \infty)$ is essentially trivial.

Theorem 1.1 (Wiener's Theorem)   *There exists a unique probability measure P on $(C, \mathcal{C})$ such that the coordinate process X on $(C, \mathcal{C}, P)$ is a standard Brownian motion.*

This P will be referred to hereafter as the *Wiener measure*. It is left as an exercise to show that a continuous process is a standard Brownian motion if and only if its distribution (see Section A.2) is the Wiener measure. When combined with Theorem 1.1, this shows that standard Brownian motion exists and is unique *in distribution*. No stronger form of uniqueness can be hoped for, because the definitive properties of standard Brownian motion refer only to the distribution of the process.

Before concluding this section we record one more important result. See Chapter 12 of Breiman (1968) for a proof.

Theorem 1.2   *If Y is a continuous process with stationary independent increments, then Y is a Brownian motion.*

This beautiful theorem shows that Brownian motion can actually be defined by stationary independent increments and path continuity alone, with normality following as a consequence of these assumptions. This may do more than any other characterization to explain the significance of Brownian motion for probabilistic modeling.

With an eye toward future requirements, we now introduce the idea of a Brownian motion *with respect to a given filtration*. Let $(\Omega, \mathcal{F}, \mathbb{F}, P)$ be a filtered probability space in the sense of Section A.1, and let X be a continuous process on this space. We say that X is a (μ, σ) Brownian motion with respect to $\mathbb{F}$, or simply a (μ, σ) Brownian motion on $(\Omega, \mathcal{F}, \mathbb{F}, P)$, if

(1.1) X is adapted,

(1.2) $X_t - X_s$ is independent of $\mathcal{F}_s$, $0 \le s \le t$, and

(1.3) X is a (μ, σ) Brownian motion in the sense of Section 1.1.

Roughly speaking, (1.1) and (1.2) say that $\mathcal{F}_t$ contains complete information about the history of X up to time t, but no information at all about the evolution of X after t. For a specific example, one may take the canonical space of Section A.3 with P the Wiener measure. In that case, X is a *standard* Brownian motion on $(\Omega, \mathcal{F}, \mathbb{F}, P)$.

1.2 Quadratic variation and local time

One of the best known properties of Brownian motion is that almost all its sample paths have infinite variation over any time interval of positive length. Thus Brownian sample paths are emphatically *not* VF functions (see Section B.2). In contrast to this negative result, a sharp positive statement can be made about the so-called quadratic variation of Brownian paths. To introduce this important concept we need a few definitions. First, a *partition* of the interval $[0, t]$ is a set of points $\Pi_t = \{t_0, t_1, \ldots, t_n\}$ with $0 = t_0 < \cdots < t_n = t$, and the *mesh* of such a partition is

$$\|\Pi_t\| := \max_{1 \le k \le n} (t_k - t_{k-1}).$$

Let $f : [0, \infty) \to \mathbb{R}$ be fixed and define

(1.4)
$$q_t(\Pi_t) := \sum_{k=1}^{n} [f(t_k) - f(t_{k-1})]^2.$$

If there exists a number q_t such that $q_t(\Pi_t) \to q_t$ as $\|\Pi_t\| \to 0$, then we call q_t the *quadratic variation* of f over $[0, t]$. The proof of the following proposition is left as an exercise.

Proposition 1.3 *If f is a continuous VF function, then $q_t = 0$ for all $t \ge 0$.*

Let X be a (μ, σ) Brownian motion on some filtered probability space $(\Omega, \mathcal{F}, \mathbb{F}, P)$, and define $Q_t(\omega)$ as q_t is defined above, but with $X(\omega)$ in place of f, assuming for the moment that the limit exists. The following proposition is proved in most standard texts.

Proposition 1.4 *For almost every $\omega \in \Omega$ we have $Q_t(\omega) = \sigma^2 t$ for all $t \ge 0$.*

Three increasingly surprising implications of Proposition 1.4 are as follows. First, the quadratic variation Q_t exists for almost all Brownian paths and all $t \ge 0$. Second, it is not zero if $t > 0$, and hence X almost surely has infinite ordinary variation over $[0, t]$ by Proposition 1.3. Finally, the quadratic variation of X does not depend on ω!

It would be difficult to overstate the significance of Proposition 1.4. We shall see later that it contains the essence of Itô's formula, and that Itô's formula is the key tool for analysis of Brownian motion and related processes. Although a complete proof of Proposition 1.4 would carry us too far afield, there are some easy calculations which at least help to make this critical result plausible. If f is replaced by X in (1.4), then the *expected value* of the sum on the right side is

$$(1.5) \quad \sum_{k=1}^{n} E\left\{[X(t_k) - X(t_{k-1})]^2\right\} = \sum_{k=1}^{n} \left[\mu^2(t_k - t_{k-1})^2 + \sigma^2(t_k - t_{k-1})\right]$$

$$\longrightarrow \sigma^2 t \qquad \text{as } \|\Pi_t\| \to 0.$$

Similarly, using the independent increments of X, one may calculate explicitly the variance of the sum. (This calculation is left as an exercise.) The variance is found to vanish as $\|\Pi_t\| \to 0$, proving that the sums converge to $\sigma^2 t$ in the L^2 sense as $n \to \infty$. Proposition 1.4 says that they also converge almost surely.

Another nice feature of Brownian paths arises in conjunction with the *occupancy measure* of the process. For each $\omega \in \Omega$ and $A \in \mathcal{B}$ (the Borel σ-algebra on $\mathbb{R}$) let

$$\cdot \quad v(t, A, \omega) := \int_0^t 1_A (X_s(\omega)) \, ds, \qquad t \geq 0,$$

with the integral defined in the Lebesgue sense. Thus $v(t, A, \cdot)$ is a random variable representing the amount of time spent by X in the set A up to time t, and $v(t, \cdot, \omega)$ is a positive measure on $(\mathbb{R}, \mathcal{B})$ having total mass t; this is the occupancy measure alluded to above. The following theorem, one of the deepest of all results relating to Brownian motion, says that the occupancy measure is absolutely continuous with respect to Lebesgue measure and has a smooth density. See Section 7.2 of Chung and Williams (1990) for a proof.

Theorem 1.5   *There exists $l : [0, \infty) \times \mathbb{R} \times \Omega \to \mathbb{R}$ such that, for almost every ω, $l(t, x, \omega)$ is jointly continuous in t and x and*

$$v(t, A, \omega) = \int_A l(t, x, \omega) \, dx \qquad \text{for all } t \geq 0 \text{ and } A \in \mathcal{B}.$$

The most difficult and surprising part of this result is the continuity of l in x, a smoothness property that testifies to the erratic behavior of Brownian paths. (Consider the occupancy measure corresponding to a continuously differentiable sample path. You will see that it does not have a continuous density at points x that are achieved as local maxima or minima of the path.) From Theorem 1.5 it follows that, for almost all ω,

$$(1.6) \qquad l(t, x, \omega) = \lim_{\epsilon \downarrow 0} \frac{1}{2\epsilon} \int_0^t 1_{[x-\epsilon, x+\epsilon]} (X_s(\omega)) \, ds$$

for all $t \geq 0$ and $x \in \mathbb{R}$. Consequently $l(\cdot, x, \omega)$ is a continuous increasing function *that increases only at time points t where* $X(t, \omega) = x$. The stochastic process $l(\cdot, x, \cdot)$ is called the *local time* of X at level x.

Proposition 1.6 *If $u : \mathbb{R} \to \mathbb{R}$ is bounded and measurable, then for almost all ω we have*

$$(1.7) \qquad \int_0^t u(X_s(\omega)) \, ds = \int_{\mathbb{R}} u(x) l(t, x, \omega) \, dx, \qquad t \geq 0.$$

Proof If u is the indicator 1_A for some $A \in \mathcal{B}$, then (1.7) follows from Theorem 1.5. Thus (1.7) holds for all simple functions u (finite linear combinations of indicators). For any positive, bounded, measurable u we can construct simple functions $\{u_n\}$ such that $u_n(x) \uparrow u(x)$ for almost every x (Lebesgue measure). Because (1.7) is valid for each u_n, it is also valid for u by the monotone convergence theorem. Moreover, the right side of (1.7) is finite because $l(t, \cdot, \omega)$ has compact support. The proof is concluded by the observation that every bounded, measurable function is the difference of two positive, bounded measurable functions. □

1.3 Strong Markov property

Again let X be a (μ, σ) Brownian motion on some filtered probability space $(\Omega, \mathcal{F}, \mathbb{F}, P)$. When we speak of stopping times (see Section A.1), implicit reference is being made to the filtration $\mathbb{F}$. Here and later we write $T < \infty$ as shorthand for the more precise statement $P\{T < \infty\} = 1$.

Theorem 1.7 *Let $T < \infty$ be a stopping time, and define $X_t^* = X_{T+t} - X_T$ for $t \geq 0$. Then X^* is a (μ, σ) Brownian motion with starting state zero and X^* is independent of $\mathcal{F}_T$.*

Let $\mathcal{F}^*$ be the smallest σ-algebra with respect to which all the random variables $\{X_t^*, t \geq 0\}$ are measurable. The last phrase of the theorem means that $\mathcal{F}_T$ and $\mathcal{F}^*$ are independent σ-algebras. Theorem 1.7 is proved in Section 37 of Billingsley (1995). This result articulates the strong Markov property in a form unique to Brownian motion. See Chapter 3 for an equivalent statement that suggests more clearly what is meant by a strong Markov process in general.

1.4 Brownian martingales

Here again we denote by X a (μ, σ) Brownian motion on a filtered probability space $(\Omega, \mathcal{F}, \mathbb{F}, P)$. Thus $X_t - X_s$ is independent of $\mathcal{F}_s$ for $s \le t$ by (1.2). If $\mu = 0$, then we have

$$(1.8) \qquad E(X_t - X_s | \mathcal{F}_s) = E(X_t - X_s) = 0$$

and

$$(1.9) \qquad E\left[(X_t - X_s)^2 | \mathcal{F}_s\right] = E\left[(X_t - X_s)^2\right] = \sigma^2(t - s).$$

Obviously (1.8) can be restated as

$$(1.10) \qquad E(X_t | \mathcal{F}_s) = X_s$$

and then the left side of (1.9) reduces to

$$
\begin{aligned}
(1.11) \qquad E\left[(X_t - X_s)^2 | \mathcal{F}_s\right] &= E(X_t^2 | \mathcal{F}_s) - 2E(X_t X_s | \mathcal{F}_s) + X_s^2 \\
&= E(X_t^2 | \mathcal{F}_s) - 2X_s E(X_t | \mathcal{F}_s) + X_s^2 \\
&= E(X_t^2 | \mathcal{F}_s) - X_s^2.
\end{aligned}
$$

Substituting (1.11) into (1.9) and rearranging terms gives

$$(1.12) \qquad E(X_t^2 - \sigma^2 t | \mathcal{F}_s) = X_s^2 - \sigma^2 s.$$

Now (1.10) and (1.12) can be restated as follows.

Proposition 1.8 *If $\mu = 0$, then X and $\{X_t^2 - \sigma^2 t, t \ge 0\}$ are martingales on $(\Omega, \mathcal{F}, \mathbb{F}, P)$.*

From (1.2) and (1.3) we know that the conditional distribution of $X_t - X_s$ given $\mathcal{F}_s$ is $N(\mu(t - s), \sigma^2(t - s))$. From this it follows that

$$(1.13) \qquad E\left[\exp\{\beta(X_t - X_s)\} | \mathcal{F}_s\right] = \exp\left\{\mu\beta(t - s) + \tfrac{1}{2}\sigma^2\beta^2(t - s)\right\}$$

for any $\beta \in \mathbb{R}$ and $s < t$. Now let

$$(1.14) \qquad q(\beta) := \mu\beta + \tfrac{1}{2}\sigma^2\beta^2, \qquad \beta \in \mathbb{R},$$

(the letter q is mnemonic for *quadratic*) and note that (1.13) can be rewritten as

$$(1.15) \qquad E\left[\exp\{\beta(X_t - X_s) - q(\beta)(t - s)\} | \mathcal{F}_s\right] = 1.$$

From (1.15) it is immediate that $E[V_\beta(t) | \mathcal{F}_s] = V_\beta(s)$, where

$$(1.16) \qquad V_\beta(t) := \exp\{\beta X_t - q(\beta)t\}, \qquad t \ge 0.$$

Thus we arrive at the following.

Proposition 1.9 V_β *is a martingale on* $(\Omega, \mathcal{F}, \mathbb{F}, P)$ *for each* $\beta \in \mathbb{R}$.

Hereafter we call V_β the *Wald martingale* with dummy variable β. It plays a central role in the calculations of Chapter 3.

1.5 Two characterizations of Brownian motion

In the calculations leading up to Proposition 1.9 we used the following: if a random variable ξ is distributed $\mathcal{N}(\mu, \sigma^2)$ then

$$E\left(e^{\beta\xi}\right) = e^{q(\beta)} \qquad \text{for all } \beta \in \mathbb{R},$$

where $q(\cdot)$ is the quadratic function (1.14). Moreover, the converse of that statement is also true: see Curtiss (1942). Combining that with the definitions in Section 1.1, one easily obtains the following converse of Proposition 1.9.

Proposition 1.10 *Let X be a continuous adapted process on a filtered probability space* $(\Omega, \mathcal{F}, \mathbb{F}, P)$. *If process* V_β *defined by* (1.16) *is a martingale for any* $\beta \in \mathbb{R}$, *then X is a* (μ, σ) *Brownian motion with respect to* $\mathbb{F}$.

The following more difficult converse is broadly useful. Its proof is beyond the scope of this book but can be found in many advanced texts; see, for example, Section 6.1 of Chung and Williams (1990).

Theorem 1.11 *Let X be a continuous martingale on a filtered probability space* $(\Omega, \mathcal{F}, \mathbb{F}, P)$, *and further suppose that, for each* $t > 0$, *X has quadratic variation t over the interval* $[0, t]$. *Then X is a standard Brownian motion with respect to* $\mathbb{F}$.

1.6 The innovation theorem

Let $W = \{W_t, t \geq 0\}$ be a standard Brownian motion defined on some probability space $(\Omega, \mathcal{F}, P)$, and let $\xi = \{\xi_t, t \geq 0\}$ be a bounded process defined on that same space, independent of W. Now imagine a decision maker who observes the process

$$(1.17) \qquad Y_t := W_t + \int_0^t \xi_s \, ds, \qquad t \geq 0,$$

and wishes to estimate, in some sense, the trajectory of ξ given the observed trajectory of Y. In this context it is usual to describe ξ as the "signal" to be estimated and W as the "noise" with which it is confounded. The decision maker seeks to "filter" the observed process Y, extracting from it an estimate of the signal ξ.

A number of such filtering problems (that is, specially structured examples of the general problem just described) will be considered in Chapter 8, where the following theorem plays a central role. In preparation, let $\mathbb{F} = \{\mathcal{F}_t, t \geq 0\}$ be the filtration generated by the observed process Y (see Section A.2 for the meaning of that phrase) and define

$$(1.18) \qquad \mu_t := E(\xi_t|\mathcal{F}_t), \qquad t \geq 0,$$

and

$$(1.19) \qquad Z_t := Y_t - \int_0^t \mu_s\,ds, \qquad t \geq 0.$$

Recall from Section A.2 that all stochastic processes are assumed to be jointly measurable in this book. Because conditional expectations are defined only up to an equivalence, (1.18) does not in itself define a *bona fide* process. That is, (1.18) does not specify a jointly measurable function $\mu_t(\omega)$, and without joint measurability the integral in (1.19) is not well defined. To remedy this problem we can invoke a basic result in stochastic process theory: there exists a (jointly measurable) process $\mu = \{\mu_t, t \geq 0\}$ such that $\mu_t = E(\xi_t|\mathcal{F}_t)$ almost surely for all $t \geq 0$. In fact, one can take μ to be what is called an optional process, thereby ensuring that the process Z defined by (1.19) is adapted to $\mathbb{F}$; see Theorem 3.6 and Lemma 3.11 of Chung and Williams (1990), or Section VI.7 of Rogers and Williams (1987). It is this choice of μ to which we refer hereafter.

In filtering theory Z is called the "innovations process," and adopting the terminology of Poor and Hadjiliadis (2008), we call the following result "the innovation theorem."

Theorem 1.12 (Innovation Theorem) *The process $Z = \{Z_t, t \geq 0\}$ defined by (1.18) and (1.19) is a standard Brownian motion with respect to the filtration $\mathbb{F}$ that is generated by Y.*

Proof From (1.18) and the tower property of conditional expectations we have that

$$(1.20) \qquad E\left(\int_s^t (\xi_u - \mu_u)\,du\Big|\mathcal{F}_s\right) = 0 \qquad \text{for } 0 \leq s \leq t.$$

Moreover,

$$(1.21) \qquad E(W_t - W_s|\mathcal{F}_s) = 0 \qquad \text{for } 0 \leq s \leq t,$$

because $W_t - W_s$ is independent of both ξ and $\{W_u,\ 0 \le u \le s\}$. Also, Z is adapted to $\mathbb{F}$, so it is a martingale with respect to $\mathbb{F}$ by (1.17), (1.19), (1.20), and (1.21). Finally, from (1.17) and (1.19) it follows that Z is continuous and has the same quadratic variation as W; that is, its quadratic variation is t over each interval $[0, t]$. Thus Z is a standard Brownian motion with respect to $\mathbb{F}$ by Theorem 1.11. $\qquad\square$

1.7 A joint distribution (Reflection principle)

Let X be a (μ, σ) Brownian motion *with starting state zero* on some filtered probability space $(\Omega, \mathcal{F}, \mathbb{F}, P)$. Also, let $M_t := \sup\{X_s, 0 \le s \le t\}$ and then define the joint distribution function

$$(1.22) \qquad F_t(x, y) := P\{X_t \le x,\ M_t \le y\}.$$

Because $X_0 = 0$ by hypothesis, one need only calculate $F_t(x, y)$ for $y \ge 0$ and $x \le y$; the discussion is hereafter restricted to (x, y) pairs satisfying those two conditions. We shall compute F for *standard* Brownian motion in this section and then extend the calculation to general μ and σ in Section 1.9. Temporarily fixing $\mu = 0$ and $\sigma = 1$, note first that

$$(1.23) \qquad \begin{aligned} F_t(x, y) &= P\{X_t \le x\} - P\{X_t \le x,\ M_t > y\} \\ &= \Phi\left(xt^{-1/2}\right) - P\{X_t \le x,\ M_t > y\} \end{aligned}$$

where $\Phi(\cdot)$ is the $\mathcal{N}(0, 1)$ distribution function. Now the term $P\{X_t \le x,\ M_t > y\}$ can be calculated heuristically using the so-called reflection principle (note that the restriction $\mu = 0$ is critical here) as follows: for every sample path of X that hits level y before time t but finishes below level x at time t, there is another equally probable path (shown by the dotted line in Figure 1.1) that hits y before time t and then travels *upward* at least $y - x$ units to finish above level $y + (y - x) = 2y - x$ at time t. Thus

$$(1.24) \qquad \begin{aligned} P\{X_t \le x,\ M_t > y\} &= P\{X_t \ge 2y - x\} \\ &= P\{X_t \le x - 2y\} = \Phi\left((x - 2y)t^{-1/2}\right). \end{aligned}$$

This argument is not rigorous, of course, but it can be made so using the strong Markov property of Section 1.3, as follows. Let T be the first t at which $X_t = y$, and define X^* as in Theorem 1.7. From Theorem 1.7 it follows that

$$\begin{aligned} P\{X_t \le x,\ M_t > y\} &= P\{T < t,\ X^*(t - T) \le x - y\} \\ &= P\{T < t,\ X^*(t - T) \ge y - x\}. \end{aligned}$$

(The strong Markov property is needed to justify the *second* of these equalities.) By definition $X^*(t - T) = X(t) - y$ and thus we arrive at (1.24). Combining (1.23) and (1.24) gives the following proposition. For the corollary, differentiate with respect to x.

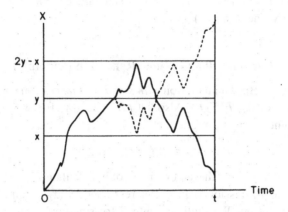

Figure 1.1 The reflection principle.

Proposition 1.13   *If $\mu = 0$ and $\sigma = 1$, then*

$$(1.25) \qquad P\{X_t \leq x,\ M_t \leq y\} = \Phi\left(xt^{-1/2}\right) - \Phi\left((x - 2y)t^{-1/2}\right).$$

Corollary 1.14 $P\{X_t \in dx,\ M_t \leq y\} = g_t(x, y)\,dx$, where

$$(1.26) \qquad g_t(x, y) := \left[\phi\left(xt^{-1/2}\right) - \phi\left((x - 2y)t^{-1/2}\right)\right] t^{-1/2}$$

and $\phi(z) := (2\pi)^{-1/2} \exp(-z^2/2)$ is the $N(0, 1)$ density function.

1.8 Change of drift as change of measure

Continuing the development in the previous section, let $T > 0$ be fixed and deterministic, and restrict X to the time domain $[0, T]$. Starting with the (μ, σ) Brownian motion $X = \{X_t,\ 0 \leq t \leq T\}$ on $(\Omega, \mathcal{F}, \mathbb{F}, P)$, suppose we want to construct a $(\mu + \theta, \sigma)$ Brownian motion, also with time domain $[0, T]$. One approach is to keep the original space $(\Omega, \mathbb{F}, P)$ and define a new process $Z_t(\omega) = X_t(\omega) + \theta t,\ 0 \leq t \leq T$. Then Z is a $(\mu + \theta, \sigma)$ Brownian motion on $(\Omega, \mathcal{F}, \mathbb{F}, P)$.

Another approach is to keep the original process X and change the probability measure. The idea is to replace P by some other probability measure

P^* such that X is a $(\mu+\theta, \sigma)$ Brownian motion on $(\Omega, \mathcal{F}, \mathbb{F}, P^*)$. In this section, we shall do just that. A positive random variable ξ will be displayed, and then P^* will be defined via

$$(1.27) \qquad P^*(A) = \int_A \xi(\omega)\, P(d\omega), \qquad A \in \mathcal{F}.$$

It is usual to express (1.27) in the more abstract form $dP^* = \xi\, dP$ and to call ξ the *density* (or Radon–Nikodym derivative) of P^* with respect to P. It will be seen that $P(\xi > 0) = 1$, so P and P^* are *equivalent measures*, meaning that $P^*(A) = 0$ if and only if $P(A) = 0$ (the two measures have the same *null sets*).

For the first two propositions below, P^* can be *any* probability measure related to P via (1.27) with $P\{\xi > 0\} = 1$. Of course, it is necessary that $E(\xi) := \int \xi\, dP = 1$, for otherwise P^* would not be a *probability* measure. It will be useful to denote by E^* the expectation operator associated with P^*, meaning that

$$(1.28) \qquad E^*(f) := \int_\Omega f\, dP^* := \int_\Omega (f \cdot \xi)\, dP := E(\xi f)$$

for measurable functions $f : \Omega \to \mathbb{R}$ such that $E|\xi f| < \infty$. Also, let $\xi(t) := E(\xi|\mathcal{F}_t)$ for $0 \le t \le T$ (see the comments immediately after (1.19) for the precise meaning of this definition), so that $\{\xi(t), 0 \le t \le T\}$ is a (strictly positive) martingale on $(\Omega, \mathcal{F}, \mathbb{F}, P)$. The proofs of the following propositions require little more than the definitions of conditional expectation and martingale, respectively; they are left as exercises.

Proposition 1.15 *Let f be a random variable with $E^*|f| < \infty$. Then $E^*(f|\mathcal{F}_t) = E(\xi f|\mathcal{F}_t)/\xi(t)$, $0 \le t \le T$.*

Corollary 1.16 *Let $Z = \{Z(t), 0 \le t \le T\}$ be a process adapted to $\mathbb{F}$. Then Z is a martingale on $(\Omega, \mathcal{F}, \mathbb{F}, P^*)$ if and only if $\{Z(t)\xi(t), 0 \le t \le T\}$ is a martingale on $(\Omega, \mathcal{F}, \mathbb{F}, P)$.*

Recall now the definitions of $q(\beta)$ and $V_\beta(t)$ from Section 1.4. Given $\theta \in \mathbb{R}$, the particular density ξ that will meet our requirements is

$$(1.29) \qquad \xi := V_\gamma(T), \qquad \text{where } \gamma := \theta/\sigma^2.$$

Before the main theorem is proved, a few observations are appropriate. First, $P\{\xi > 0\} = 1$. Second, V_γ is a martingale on $(\Omega, \mathcal{F}, \mathbb{F}, P)$ by Proposition 1.9, so

$$(1.30) \qquad E(\xi|\mathcal{F}_0) = E\left[V_\gamma(T)|\mathcal{F}_0\right] = V_\gamma(0) = 1,$$

implying $E(\xi) = 1$ as required. (The last equality in (1.30) follows from the fact that $X_0 = 0$ by assumption.) More generally,

(1.31) $E(\xi|\mathcal{F}_t) = E\left[V_\gamma(T)|\mathcal{F}_t\right] = V_\gamma(t), \qquad 0 \le t \le T.$

Comparing (1.31) with the earlier definition of $\xi(t)$, we see that, when ξ is defined by (1.29),

(1.32) $\xi(t) = V_\gamma(t), \qquad 0 \le t \le T.$

Theorem 1.17 (Change of Measure Theorem) *Given $\theta \in \mathbb{R}$, let ξ and P^* be defined by (1.29) and (1.27), respectively. Then X is a $(\mu + \theta, \sigma)$ Brownian motion on $(\Omega, \mathcal{F}, \mathbb{F}, P^*)$.*

Proof   Let $\beta \in \mathbb{R}$ be arbitrary and define

(1.33) $V_\beta^*(t) := \exp\left[\beta X_t - q^*(\beta)t\right], \qquad 0 \le t \le T,$

where

(1.34) $q^*(\beta) := (\mu + \theta)\beta + \frac{1}{2}\sigma^2\beta^2.$

By Proposition 1.10 it suffices to show that

(1.35) V_β^* is a martingale on $(\Omega, \mathcal{F}, \mathbb{F}, P^*)$.

Now Corollary 1.16 shows that (1.35) is equivalent to the following:

(1.36) $\left\{V_\beta^*(t)\xi(t),\ 0 \le t \le T\right\}$ is a martingale on $(\Omega, \mathcal{F}, \mathbb{F}, P)$.

From (1.33) and (1.32) we have that

(1.37)
$$V_\beta^*(t)\xi(t) = \exp\left[\beta X_t - q^*(\beta)t\right]\exp\left[\gamma X_t - q(\gamma)t\right]$$
$$= \exp\left[(\beta + \gamma)X_t - \psi(\beta)t\right]$$

where

(1.38) $\psi(\beta) := q^*(\beta) + q(\gamma) = \left[(\mu + \theta)\beta + \frac{1}{2}\sigma^2\beta^2\right] + \left[\mu\gamma + \frac{1}{2}\sigma^2\gamma^2\right].$

Using the fact that $\gamma := \theta/\sigma^2$, readers can now verify that $\psi(\beta) = \mu(\beta + \gamma) + \sigma^2(\beta + \gamma)^2/2 := q(\beta + \gamma)$. Thus (1.37) says that $V_\beta^*(t)\xi(t) = V_{\beta+\gamma}(t)$. Finally, $V_{\beta+\gamma}$ is a martingale on $(\Omega, \mathcal{F}, \mathbb{F}, P)$ by Proposition 1.9, so (1.36) holds and the proof is complete. $\square$

It should be emphasized that the change of measure theorem is only valid when one views X as a process with finite time horizon. However, it can be generalized to the case where T is a stopping time, and many other generalizations are known. See Section 3.5 of Karatzas and Shreve (1998) for a taste of the more general theory.

1.9 A hitting time distribution

Returning to the analysis begun in Section 1.7, we now use the change of measure theorem to calculate the joint distribution of X_t and M_t in general.

Proposition 1.18 *For general values of μ and σ we have*

$$(1.39) \qquad P\{X_t \in dx, \ M_t \le y\} = f_t(x, y)\, dx$$

where

$$(1.40) \qquad f_t(x, y) = (1/\sigma) \exp(\mu x/\sigma^2 - \mu^2 t/2\sigma^2) g_t(x/\sigma, y/\sigma)$$

and $g_t(\cdot, \cdot)$ is defined by (1.26).

Proof Only the case $\sigma = 1$ will be treated here; the extension to general σ is accomplished by a straightforward rescaling. Suppose initially that X is a *standard* Brownian motion on $(\Omega, \mathcal{F}, \mathbb{F}, P)$ so that

$$(1.41) \qquad P\{X_t \in dx, \ M_t \le y\} = g_t(x, y)\, dx$$

by Corollary 1.14. Now fix $t > 0$, let $\mu \in \mathbb{R}$ be arbitrary, set

$$(1.42) \qquad \xi := \exp\left(\mu X_t - \tfrac{1}{2}\mu^2 t\right)$$

and define a new probability measure P^* by taking $dP^* = \xi\, dP$. The change of measure theorem 1.17 says that $\{X_s, 0 \le s \le t\}$ is a $(\mu, 1)$ Brownian motion under P^*, so the desired result (specialized to $\sigma = 1$) is equivalently stated as

$$(1.43) \qquad P^*\{X_t \in dx, \ M_t \le y\} = e^{\mu x - \mu^2 t/2} g_t(x, y)\, dx.$$

To simplify typography in the proof of (1.43), let us denote by $1(A)$ the random variable that has value 1 on A and value zero otherwise. Using (1.41),

$$
\begin{aligned}
P^*\{X_t \le x, \ M_t \le y\} &= E^* \left[1(X_t \le x, \ M_t \le y)\right] \\
&= E\left[\xi \cdot 1(X_t \le x, \ M_t \le y)\right] \\
&= E\left[e^{\mu X_t - \mu^2 t/2} 1(X_t \le x, \ M_t \le y)\right] \\
&= \int_{-\infty}^{x} e^{\mu z - \mu^2 t/2} P\{X_t \in dz, \ M_t \le y\} \\
&= \int_{-\infty}^{x} e^{\mu z - \mu^2 t/2} g_t(z, y)\, dz.
\end{aligned}
$$

Differentiating with respect to x gives (1.43) as required. $\qquad\square$

Corollary 1.19   *Let $F_t(x, y) := P\{X_t \leq x, \ M_t \leq y\}$ as in Section 1.7. For general values of μ and σ we have*

(1.44) $$F_t(x, y) = \Phi\left(\frac{x - \mu t}{\sigma t^{1/2}}\right) - e^{2\mu y/\sigma^2}\Phi\left(\frac{x - 2y - \mu t}{\sigma t^{1/2}}\right).$$

Proof   Again we treat only the case $\sigma = 1$. By specializing the general formula (1.40) for f accordingly, we obtain

$$F_t(x, y) = \int_{-\infty}^{x} f_t(z, y)\, dz$$

$$= e^{-\mu^2 t/2} \int_{-\infty}^{x} e^{\mu z} t^{-1/2}\left[\phi(t^{-1/2}z) - \phi\left(t^{-1/2}(z - 2y)\right)\right] dz.$$

Elementary manipulations now reduce this expression to

(1.45) $$F_t(x, y) = e^{\mu x - \mu^2 t/2}\left[\Psi(x) - \Psi(x - 2y)\right]$$

where

(1.46) $$\Psi(x) := \int_{-\infty}^{0} t^{-1/2} e^{\mu z}\phi\left(t^{-1/2}(z + x)\right) dz.$$

Now let $h(x, t) := t^{-1/2}(x - \mu t)$. Writing out $\phi(\cdot)$ and completing the square in the exponent, we have

$$\Psi(x)$$

$$= \int_{-\infty}^{0} (2\pi t)^{-1/2} \exp\left\{\mu z - \frac{(z^2 + 2xz + x^2)}{2t}\right\} dz$$

$$= \int_{-\infty}^{0} (2\pi t)^{-1/2} \exp\left\{-\left[\frac{z^2 + 2(x - \mu t)z + (x - \mu t)^2}{2t}\right] + \left(\frac{\mu^2 t}{2} - \mu x\right)\right\} dz$$

$$= e^{-\mu x + \mu^2 t/2} \int_{-\infty}^{0} (2\pi t)^{-1/2} \exp\left\{-\frac{1}{2}\left[t^{-1/2}(z + x - \mu t)\right]^2\right\} dz$$

$$= e^{-\mu x + \mu^2 t/2} \int_{-\infty}^{h(x,t)} \phi(u)\, du$$

$$= e^{-\mu x + \mu^2 t/2}\Phi(h(x, t)).$$

Substituting this into (1.45) gives the desired formula.   □

If we define $T(y)$ as the first t at which $X_t = y$ (possibly $+\infty$ if $\mu < 0$), then obviously $T(y) > t$ if and only if $M_t < y$. Letting $x \uparrow y$ in (1.44) gives

$$P\{T(y) > t\} = P\{M_t < y\} = F_t(y, y)$$

(1.47)

$$= \Phi\left(\frac{y - \mu t}{\sigma t^{1/2}}\right) - e^{2\mu y/\sigma^2}\Phi\left(\frac{-y - \mu t}{\sigma t^{1/2}}\right)$$

for $y > 0$. With this we have calculated explicitly the *one-sided first passage time distribution for Brownian motion with drift*.

1.10 Reflected Brownian motion

We continue to denote by X a (μ, σ) Brownian motion with $X_0 = 0$. Let us now define an increasing process L and a positive process Z by setting

$$(1.48) \qquad L_t := -\inf_{0 \le s \le t} X_s, \qquad\qquad t \ge 0,$$

and

$$(1.49) \qquad Z_t := X_t + L_t = \sup_{0 \le s \le t}(X_t - X_s), \quad t \ge 0.$$

Later Z will be called *reflected Brownian motion* (RBM), or Brownian motion modified by a *lower reflecting barrier* at zero. (As noted in the Introduction, this is potentially confusing terminology, but it is now fixed by long use.) The simple representation (1.49) is specific to the case $X_0 = 0$, but in Section 3.7 a general representation for an arbitrary starting state will be developed. The probabilistic and the analytic theory of RBM will be developed in later chapters.

A slight modification of the arguments used in Section 1.7 and Section 1.9 gives the joint distribution of X_t and L_t, from which one can obviously calculate the distribution of Z_t. But here is an easier way. Fix $t > 0$ and for $0 \le s \le t$ let $X_s^* = X_t - X_{t-s}$. Note that $X^* := \{X_s^*, 0 \le s \le t\}$ has stationary, independent increments with $X_0^* = 0$ and $X_s^* \sim \mathcal{N}(\mu s, \sigma^2 s)$. Thus X^* is another (μ, σ) Brownian motion with starting state zero. Combining this with (1.49) we get

$$(1.50) \qquad \begin{aligned} Z_t &= \sup_{0 \le s \le t}(X_t - X_s) = \sup_{0 \le s \le t}(X_t - X_{t-s}) \\ &= \sup_{0 \le s \le t} X_s^* \sim \sup_{0 \le s \le t} X_s := M_t. \end{aligned}$$

(Here the symbol $\sim$ denotes equality in distribution.) Thus the distributions of Z_t and M_t coincide for each fixed t, although the distributions of the complete processes Z and M are very different. (For example, M has increasing sample paths, but Z does not.)

The marginal distribution of M was displayed earlier in (1.47). Combining this with (1.50) gives

$$(1.51) \qquad P\{Z_t \le z\} = \Phi\left(\frac{z - \mu t}{\sigma t^{1/2}}\right) - e^{2\mu z/\sigma^2} \Phi\left(\frac{-z - \mu t}{\sigma t^{1/2}}\right)$$

for all $t \geq 0$. Thus as $t \to \infty$,

(1.52) $P\{Z_t \leq z\} \longrightarrow \begin{cases} 1 - e^{2\mu z/\sigma^2} & \text{if } \mu < 0 \\ 0 & \text{if } \mu \geq 0. \end{cases}$

For $\mu < 0$ the limit distribution (1.52) is exponential with mean $\sigma^2/2|\mu|$.

We shall continue the analysis of Z later using the machinery of stochastic calculus. To prepare the way, it will be useful to record some properties of the process L. (Everything said here would apply equally well to M.) It is obviously continuous, but the following well-known proposition shows that L increases in a highly irregular fashion.

Proposition 1.20   *For almost every ω and any $t > 0$, $L(\omega)$ has uncountably many points of increase in $[0, t]$, but the set of all such points has (Lebesgue) measure zero.*

Because the sample paths of L increase only on a set of measure zero, they cannot be absolutely continuous; L cannot be expressed as the integral of another process (see Section B.1). One cannot speak of the *rate* at which L increases, although its sample paths are continuous and increasing and are therefore VF functions (see Section B.2). Because L plays such an important role in this book, the distinction between VF processes and absolutely continuous processes is an important one for us.

1.11 Problems and complements

Problem 1.1   Let X be a continuous process with distribution Q (see Section A.2). It was stated in Section 1.1 that the definitive properties of standard Brownian motion involve only the distribution of the process. This means that X is a standard Brownian motion if and only if Q satisfies certain conditions. Write out in precise mathematical form what those conditions are, and then show that X is a standard Brownian motion if and only if Q is the Wiener measure. Although this problem requires nothing more than shuffling definitions, it is difficult for those who have never dealt with stochastic processes in abstract terms. It requires that one understand the general distinction between a stochastic process and its distribution, and the specific distinction between standard Brownian motion and the Wiener measure.

Problem 1.2   Prove Proposition 1.3, which says that a continuous VF function has zero quadratic variation.

Problem 1.3 Calculate the variance of the sum on the left side of (1.5) and show that this vanishes as $n \to \infty$.

Problem 1.4 Let X be the coordinate process on C as in Section A.3 and let $v(t, A, \omega)$ be the occupancy measure for X, as in Section 1.2. Consider the particular point $\omega \in C$ defined by $\omega(t) = (1 - t)^2$, $t \ge 0$. Fix a time $t > 1$ and describe $v(t, \cdot, \omega)$ in precise mathematical terms. Observe that this measure on $(\mathbb{R}, \mathcal{B})$ is absolutely continuous (with respect to Lebesgue measure) but its density is not continuous. This substantiates a claim made in Section 1.2.

Problem 1.5 Prove Proposition 1.15 and Corollary 1.16. This is just a matter of verification, using the definitions of conditional expectation and martingale.

2

Stochastic Storage Models

Consider a firm that produces a single durable commodity on a make-to-stock basis. Production flows into a finished goods inventory, and demand that cannot be met from stock on hand is simply lost, with no adverse effect on future demand. The price of the output good is fixed, and demand is viewed as an exogenous source of uncertainty. Similarly, we consider plant, equipment, and workforce size to be fixed for now, but there may be uncertainty about actual production quantities because of mechanical failures, worker absenteeism, and so forth. This firm and its market, portrayed schematically in Figure 2.1, constitute a simple storage system. It consists of an input process (production), an output process (demand), and an intermediate buffer storage (the finished goods inventory) that serves to decouple input and output. Many mathematical models of such systems have been developed in the literature of operations research, most often under the rubric of "queuing theory."

Figure 2.1 A simple storage system.

The abstract language of input processes, output processes, and storage buffers will be used hereafter, but the content of the buffer will be called inventory, and all our examples involve production systems. In this chapter we develop a crude model of buffered flow, making no attempt to portray physical structure beyond that apparent in Figure 2.1. Actually, two models will be advanced, one with infinite buffer capacity and one with finite capacity. In each case, system flows are represented by continuous stochastic processes. Thus our models have little direct relevance to systems where individual inventory items are physically or economically significant, but for discrete item systems with high-volume flow, the continuity assumption may be viewed as a convenient and harmless idealization.

2.1 Buffered stochastic flow

Assume that the buffer in Figure 2.1 has infinite capacity. To model the system, we take as primitive a constant $X_0 \geq 0$ and two increasing, continuous stochastic processes $A = \{A_t, t \geq 0\}$ and $B = \{B_t, t \geq 0\}$ with $A_0 = B_0 = 0$. Interpret X_0 as the initial inventory level, A_t as the cumulative input up to time t, and B_t as the cumulative *potential* output up to time t. In other words, B_t is the total output that can be realized over the time interval $[0, t]$ *if the buffer is never empty*; more generally, $B_t - B_s$ is the maximum possible output over the interval $(s, t]$. If emptiness does occur, then some of this potential output will be lost. We denote by L_t the amount of potential output lost up to time t because of such emptiness, so *actual* output over $[0, t]$ is $B_t - L_t$. Setting

$$(2.1) \qquad\qquad X_t := X_0 + A_t - B_t$$

the inventory at time t is then given by

$$(2.2) \qquad\qquad Z_t := X_0 + A_t - (B_t - L_t) = X_t + L_t.$$

Most of our attention will focus on this *inventory process* $Z = \{Z_t, t \geq 0\}$. It remains to define the lost potential output process L in terms of primitive model elements, and for that we simply assume (or require) that

$$(2.3) \qquad L \text{ is increasing and continuous with } L_0 = 0 \text{ and}$$

$$(2.4) \qquad L \text{ increases only when } Z = 0.$$

Conditions (2.3) and (2.4) together say that output is (by assumption) sacrificed in the minimum amounts consistent with the physical restriction

$$(2.5) \qquad\qquad Z_t \geq 0 \qquad \text{for all } t \geq 0.$$

In the next section it will be shown that conditions (2.2) to (2.5) uniquely determine L and further imply the concise representation

$$(2.6) \qquad\qquad L_t = \sup_{0 \leq s \leq t} X_s^-.$$

Because X is defined in terms of primitive elements by (2.1), this completes the precise mathematical specification of our stochastic storage model with infinite buffer capacity.

A critical feature of this construction is that L and Z depend on A and

B only through their difference, so one may view X as the sole primitive element of our system model. Hereafter we shall refer to X as a *netflow process*. This same term will be used later in other contexts, always to describe a net of potential input minus potential output. The development above requires that X have continuous sample paths, but thus far no probabilistic assumptions have been imposed. The emphasis in this chapter is on construction of sample paths rather than on probabilistic analysis.

2.2 The one-sided reflection mapping

Let $C := C[0, \infty)$ as in Section A.2. Elements of C will often be called paths or trajectories rather than functions, and the generic element of C will be denoted by $x = (x_t, t \geq 0)$. We now define mappings $\psi, \phi : C \to C$ by setting

$$(2.7) \qquad \psi_t(x) := \sup_{0 \leq s \leq t} x_s^- \qquad \text{for } t \geq 0$$

and

$$(2.8) \qquad \phi_t(x) := x_t + \psi_t(x) \qquad \text{for } t \geq 0.$$

For purposes of discussion, fix $x \in C$ and let $l := \psi(x)$ and $z := \phi(x) = x + l$. We say that z is obtained from x by imposition of a *lower reflecting barrier* at zero. The pair (ψ, ϕ) will be called the *one-sided reflection mapping* with lower barrier at zero. The effect of this path-to-path transformation is shown graphically in Figure 2.2, where the dotted line is $-l_t$. Note that $l = 0$ and hence $z = x$ up until the first time t at which $x_t = 0$. Thereafter z_t equals the amount by which x_t exceeds the minimum value of x over $[0, t]$.

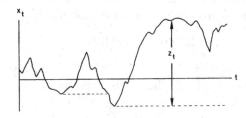

Figure 2.2 The one-sided reflection mapping.

The following proposition and proof are due to El Karoui and Chaleyat-Maurel (1978).

Proposition 2.1 *Suppose $x \in C$ and $x_0 \geq 0$. Then $\psi(x)$ is the unique function l such that*

(2.9)   *l is continuous and increasing with $l_0 = 0$,*

(2.10)   *$z_t := x_t + l_t \geq 0$ for all $t \geq 0$, and*

(2.11)   *l increases only when $z = 0$.*

Remark 2.2   Let l be any function on $[0, \infty)$ satisfying (2.9) and (2.10) alone. It is easy to show that $l_t \geq \psi_t(x)$ for all $t \geq 0$. In this sense, the *least* solution of (2.9) and (2.10) alone is obtained by taking $l = \psi(x)$.

Proof   Fix $x \in C$ and set $l := \psi(x)$ and $z := x + l$. It is left as an exercise to show that this l does in fact satisfy (2.9) to (2.11). To prove uniqueness, let l^* be any other solution of (2.9) to (2.11) and set $z^* := x + l^*$. Setting $y := z^* - z = l^* - l$, we note that y is a continuous VF function with $y_0 = 0$. Thus the Riemann–Stieltjes chain rule (B.2) gives

$$f(y_t) = f(0) + \int_0^t f'(y)\,dy \tag{2.12}$$

for any continuously differentiable $f : \mathbb{R} \to \mathbb{R}$. Taking $f(y) = y^2/2$, we see that (2.12) reduces to

$$\frac{1}{2}(z_t^* - z_t)^2 = \int_0^t (z^* - z)\,dl^* + \int_0^t (z - z^*)\,dl. \tag{2.13}$$

We know that l^* increases only when $z^* = 0$, and $z \geq 0$, so the first term on the right side of (2.13) is ≤ 0, and identical reasoning shows that the second term is ≤ 0 as well. But because the left side is ≥ 0, both sides must be zero. This shows that $z^* = z$ and hence $l^* = l$, and the proof is complete.   □

Note that the property $l_0 = 0$ in (2.9) depends on the assumption that $x_0 \geq 0$. The following proposition shows that the one-sided reflection mapping has a certain *memoryless property*. It will be used later to prove the strong Markov property of reflected Brownian motion.

Proposition 2.3   *Fix $x \in C$ and set $l := \psi(x)$ and $z := \phi(x) = x + l$. Fix $T > 0$ and define $x_t^* = z_T + (x_{T+t} - x_T)$, $l_t^* = l_{T+t} - l_T$, and $z_t^* = z_{T+t}$ for $t \geq 0$. Then $l^* = \psi(x^*)$ and $z^* = \phi(x^*)$.*

Because the proof of Proposition 2.3 is just a matter of verification, it is left as an exercise. Pursuant to the observation in Remark 2.2, it is often helpful to think of l_t as the cumulative amount of *control* exerted by an observer of the sample path x up to time t. This observer must increase l

fast enough to keep $z := x + l$ positive but wishes to exert as little control as possible subject to that constraint.

2.3 Finite buffer capacity

Consider again the simple storage system of Section 2.1, assuming now that the buffer has finite capacity b. Except as noted below, the assumptions and notation of Section 2.1 remain in force. In particular, the system netflow process is defined by $X_t := X_0 + A_t - B_t$, and L_t denotes the amount of potential output lost up to time t due to emptiness of the buffer. In the current context one must interpret A as a *potential* input process; some of this potential input may be lost when the buffer is full. For reasons that will become clear in the next section, we denote by U_t the total amount of potential input lost up to time t. Thus actual input up to time t is $A_t - U_t$, and the inventory process Z is given by

$$(2.14) \qquad Z_t = X_0 + (A_t - U_t) - (B_t - L_t) = X_t + L_t - U_t.$$

Now how are L and U to be defined in terms of the primitive model elements? Assuming that $X_0 \in [0, b]$, it is more or less obvious from the development in Section 2.1 and Section 2.2 that L and U should be uniquely determined by the following properties:

$(2.15) \qquad L$ and U are continuous and increasing with $L_0 = U_0 = 0$,

$(2.16) \qquad Z_t := (X_t + L_t - U_t) \in [0, b]$ for all $t \geq 0$, and

$(2.17) \qquad L$ and U increase only when $Z = 0$ and $Z = b$, respectively.

In the next section it will be shown that (2.15) to (2.17) do in fact determine L and U uniquely. Explicit formulas for the processes L and U are given by Kruk et al. (2007), but those formulas will not be used here. A crucial point is that L, U, and Z depend on primitive model elements only through the netflow process X.

A finite buffer may represent either a physical restriction on storage space or a policy restriction that shuts off input when buffer stock reaches a certain level. In the context of production systems, input is almost always controllable, and it is simply irrational to let inventory levels fluctuate without restriction. Thus the model described here is fundamentally more interesting than the one developed in Section 2.1.

2.4 The two-sided reflection mapping

Fix $b > 0$ and let C^* be the set of all functions $x \in C$ such that $x_0 \in [0, b]$. Given $x \in C^*$, we would like to find a pair of functions (l, u) such that

(2.18) l and u are increasing and continuous with $l_0 = u_0 = 0$,

(2.19) $z_t := (x_t + l_t - u_t) \in [0, b]$ for all $t \geq 0$, and

(2.20) l and u increase only when $z = 0$ and $z = b$, respectively.

Note that (2.20) associates l and u with the *lower* barrier at zero and *upper* barrier at b, respectively. If we consider u to be given, then the requirements imposed on l by (2.18) to (2.20) are those that define a lower reflecting barrier at zero. That is, (2.18) to (2.20) and Proposition 2.1 together imply that

(2.21) $$l_t = \psi_t(x - u) := \sup_{0 \leq s \leq t} (x_s - u_s)^-.$$

In exactly the same way, u may be expressed in terms of l via

(2.22) $$u_t = \psi_t(b - x - l) := \sup_{0 \leq s \leq t} (b - x_s - l_s)^-.$$

It will now be proved that (2.21) and (2.22) together uniquely determine l and u. The function z defined by (2.19) may be pictured as in Figure 2.3, where the lower dotted line is $u_t - l_t$ and the upper dotted line is $b + u_t - l_t$. We shall henceforth say that z is obtained from x through imposition of a *lower reflecting barrier at zero and an upper reflecting barrier at b*.

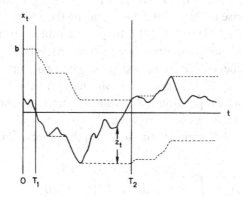

Figure 2.3 The two-sided reflection mapping.

Proposition 2.4   *For each $x \in C^*$, there is a unique pair of continuous functions (l, u) satisfying (2.21) and (2.22), and this same pair uniquely satisfies (2.18) to (2.20).*

Definition 2.5   We define mappings $f, g, h : C^* \to C$ by setting $f(x) :=$ l, $g(x) := u$, and $h(x) := x + l - u$. Hereafter (f, g, h) will be called the *two-sided reflection mapping* with lower barrier at zero and upper barrier at b.

Proof   We first construct a solution of (2.21) and (2.22) by successive approximations. Beginning with the trial solution $l_t^0 = u_t^0 = 0 \ (t \geq 0)$, we set

$$(2.23) \qquad l_t^{n+1} := \psi_t(x - u^n) := \sup_{0 \leq s \leq t} (x_s - u_s^n)^-$$

and

$$(2.24) \qquad u_t^{n+1} := \psi_t(b - x - l^n) := \sup_{0 \leq s \leq t} (b - x_s - l_s^n)^-$$

for $n = 0, 1, \ldots$ and $t \geq 0$. Observe that $l_t^1 \geq l_t^0$ and $u_t^1 \geq u_t^0$ for all t, and hence (by induction) that l_t^n and u_t^n are increasing in n for each fixed t. Thus we have

$$(2.25) \qquad l_t^n \uparrow l_t \quad \text{and} \quad u_t^n \uparrow u_t \qquad \text{as } n \uparrow \infty.$$

Furthermore, it is easy to show that the convergence is achieved in a finite number of iterations for each fixed t, and the requisite number of iterations is an increasing function of t. For example, in Figure 2.3 we have $l_t = l_t^0$ and $u_t = u_t^0$ if $0 \leq t \leq T_1$, $l_t = l_t^1$ and $u_t = u_t^1$ if $T_1 \leq t \leq T_2$, and so forth. (It is left as an exercise to show that $T_n \to \infty$, using the assumed continuity of x.) From this and (2.23) and (2.24) it follows that the limit functions l and u are finite-valued, are continuous, and jointly satisfy (2.21) and (2.22).

To prove uniqueness, let (l, u) and (l^*, u^*) be two pairs of continuous functions satisfying (2.21) and (2.22), and let $z := x + l - u$ and $z^* := x + l^* - u^*$. From Proposition 2.1 it follows that (l, u) and (l^*, u^*) both satisfy (2.18) to (2.20) as well. Now let $y := z^* - z = (l^* - l) - (u^* - u)$. Using the Riemann–Stieltjes chain rule as in the proof of Proposition 2.1, we find that

$$(2.26) \qquad \frac{1}{2}(z_t^* - z_t)^2 = \int_0^t (z^* - z) \, dl^* + \int_0^t (z - z^*) \, dl$$

$$+ \int_0^t (z - z^*) \, du^* + \int_0^t (z^* - z) \, du.$$

Also as in the proof of Proposition 2.1, we use (2.18) to (2.20) to conclude that each term on the right side of (2.26) is ≤ 0, whereas the left side is ≥ 0, and hence each side is zero. Thus $z^* = z$, from which it follows easily that $l^* = l$ and $u^* = u$. Thus there is exactly one continuous pair (l, u) satisfying (2.21) and (2.22). As we observed earlier, (2.18) to (2.20) are equivalent to (2.21) and (2.22) for continuous pairs (l, u) by Proposition 2.1, and this proves the last statement of the proposition. □

Corollary 2.6 *For each fixed t, both* $l_t := f_t(x)$ *and* $u_t := g_t(x)$ *depend on x only through* $(x_s, 0 \leq s \leq t)$.

Proof Immediate from the construction (2.23) to (2.25). □

Proposition 2.7 *Fix* $x \in C$ *and let* $l := f(x)$, $u := g(x)$, *and* $z := h(x)$ *as above. Fix* $T > 0$ *and define* $x_t^* := z_T + (x_{T+t} - x_T)$, $l_t^* := l_{T+t} - l_T$, $u_t^* := u_{T+t} - u_T$, *and* $z_t^* := z_{T+t}$ *for* $t \geq 0$. *Then* $l^* = f(x^*)$, $u^* = g(x^*)$, *and* $z^* = h(x^*)$.

Proof Starting with the fact that x, l, u, z all satisfy (2.18) to (2.20), it is easy to verify that x^*, l^*, u^*, z^* satisfy these same relations. The second uniqueness statement of Proposition 2.4 then establishes the desired proposition. □

2.5 Measuring system performance

In the design and operation of storage systems, one is typically concerned with a tradeoff between system throughput characteristics and the costs associated with inventory. Generally speaking, one can decrease the amount of lost potential input and output (which amounts to improving capacity utilization) by tolerating larger buffer stocks, but such stocks are costly in their own right.

To put the discussion on a concrete footing, consider again the single-product firm described at the beginning of this chapter. Recall that production flows into a finished goods inventory, and demand that cannot be met from stock on hand is simply lost with no adverse effect on future demand. Let π denote the selling price (in dollars per unit of production) and let B_t denote total demand over the time interval $[0, t]$. The latter notation is chosen for consistency with previous use in Section 2.1 and Section 2.3.

Assuming plant and equipment are fixed, suppose that the firm must select at time zero a workforce size, or equivalently a regular-time production capacity. For simplicity, assume that the workforce size cannot be varied thereafter, the firm being obliged to pay workers their regular wages regardless of whether they are productively employed. Let k be the capacity level selected, in units of production per time unit. The firm then incurs a labor cost of wk dollars per time unit ever afterward, where $w > 0$ is a specified wage rate, even if it occasionally chooses to operate below capacity. For current purposes, overtime production is assumed to be impossible (see Problem 2.8). In addition to its labor costs, the firm incurs a materials cost of m dollars per unit of *actual* production. Given the initial capacity decision (workforce level), labor costs are fixed, and thus the *marginal cost of production is m dollars per unit*. A physical holding cost of p dollars is incurred per time unit for each unit of production held in inventory. This includes such costs as insurance and security; it does *not* include the financial cost of holding inventory. (By financial cost we mean the opportunity loss on money tied up in inventory. More will be said on this subject shortly.)

It is assumed that the firm earns interest at rate $\lambda > 0$, compounded continuously, on funds that are not required for production operations. Continuous compounding means that one dollar invested at time zero returns $\exp(\lambda t)$ dollars of principal plus interest at time t. Thus a cost or revenue of one dollar at time t is equivalent in value to a cost or revenue of $\exp(-\lambda t)$ dollars at time zero. Finally, we assume that the cumulative demand process B satisfies

(2.27) $E(B_t) = at$ for all $t \geq 0$ $(a > 0)$ and

(2.28) $e^{-\lambda t} B_t \to 0$ almost surely as $t \to \infty$.

For one specific demand model that satisfies (2.27) and (2.28), we may suppose that the time axis can be divided into periods of unit length, that demand increments during successive periods form a sequence of independent and identically distributed random variables with mean a and finite variance, and that demand arrives at a constant rate during each period. For this *linearized random walk* model of demand, property (2.27) is obvious and (2.28) follows from the strong law of large numbers. (The proof of this statement is left as an exercise.)

The firm must choose a capacity level k at time zero and then at each time $t \geq 0$ select a production rate from the interval $[0, k]$. When a production rate below k is selected, we shall say that *undertime* is being employed. For purposes of initial discussion, let us assume that management follows a *single-barrier policy* for production control after time zero. This means that production continues at the capacity rate k until inventory hits some chosen level $b > 0$, and then undertime is employed in the minimum amounts necessary to keep inventory at or below level b. With this policy, our make-to-stock production system is a storage system with finite buffer capacity (see Section 2.3); the potential input process is $A_t := kt$, and potential output is given by the demand process B. In the current context, Z_t represents the finished goods inventory level at time t, L_t is the cumulative demand lost up to time t, and U_t is the cumulative undertime worked (potential production foregone) up to time t.

The firm's objective is to maximize the expected present value of sales revenues received minus operating expenses incurred over an infinite planning horizon, where discounting is continuous at interest rate λ. The actual production and sales volumes up to time t are given by $kt - U_t$ and $B_t - L_t$, respectively; thus the stated objective amounts to maximization of

$$(2.29) \quad V := E\left[\pi \int_0^\infty e^{-\lambda t}(dB - dL) - wk \int_0^\infty e^{-\lambda t}\, dt \right.$$
$$\left. -m \int_0^\infty e^{-\lambda t}(k\, dt - dU) - p \int_0^\infty e^{-\lambda t} Z_t\, dt\right],$$

where the integrals involving dB, dL, and dU are defined path by path in the Riemann–Stieltjes sense (see Appendix B). The first term inside the expectation in (2.29) represents the present value of sales revenues, the second is the present value of labor costs, the third term is the present value of material costs (incremental production costs), and the last is the present value of inventory holding costs. It should be emphasized that the opportunity loss on capital tied up in inventory is fully accounted for by the discounting in (2.29); therefore p should include only out-of-pocket expenses associated with holding inventory. To put it another way, no explicit financial cost of holding inventory appears in (2.29) and including such a cost would be double counting. In a moment, however, we shall derive an equivalent measure of system performance in which a financial cost of inventory *does* appear. Readers who are not familiar with present value manipulations, and skeptical as to the appropriateness of (2.29) as a performance measure, may wish to consult Section 7.9. There it is shown that maximizing a discounted measure like V is equivalent to maximizing the firm's expected total assets at a distant time of reckoning.

It will now be shown that maximizing V is equivalent to minimizing another, somewhat simpler, performance measure. As a first step, consider the ideal situation where $B_t = at$ for all $t \geq 0$, meaning that demand arrives deterministically at constant rate a. We shall assume that

$$(2.30) \qquad\qquad \pi - w - m > 0,$$

for otherwise the system optimization problem would be uninteresting. (If $\pi - w - m \leq 0$, it is best to set $k = 0$ and go out of business.) With deterministic demand, one would choose $k = a$, of course, meaning that units are produced precisely as demanded, labor and materials are paid for only as required for such production, and no inventory is held. The corresponding *ideal profit level* (in present value terms) would be

$$(2.31) \qquad I := \int_0^\infty e^{-\lambda t}(\pi - w - m)a \, dt = \frac{(\pi - w - m)a}{\lambda}.$$

Now actual system performance under an arbitrary operating policy will be measured incrementally from this ideal. First, let

$$(2.32) \qquad \mu := k - a,$$

$$(2.33) \qquad \delta := \pi - m, \quad \text{and} \quad h := p + m\lambda.$$

We call μ the *excess capacity*; it is the amount (possibly negative) by which chosen capacity exceeds the average demand rate. Interpret δ as a *contribution margin*; once the capacity level is fixed, each unit of sales contributes δ dollars to profit and the coverage of fixed costs. Finally, h may be viewed as the *effective cost of holding inventory*; it consists of physical holding costs plus an opportunity loss rate of λ times the marginal production cost m. It is assumed hereafter that $Z_0 = 0$.

Proposition 2.8 $V = I - \Delta$, *where*

$$(2.34) \qquad \Delta := E\left[\int_0^\infty e^{-\lambda t}(\delta \, dL + w\mu \, dt + hZ_t \, dt)\right].$$

Remark 2.9   Because demand is exogenous, I is an uncontrollable constant, and thus our original objective of maximizing V is equivalent to minimizing Δ.

Proof   From (2.27) and (2.28) it follows that

$$(2.35) \qquad \begin{aligned} E\left(\int_0^\infty e^{-\lambda t} \, dB\right) &= E\left(\int_0^\infty \lambda e^{-\lambda t} B_t \, dt\right) \\ &= \int_0^\infty \lambda e^{-\lambda t} at \, dt = \int_0^\infty e^{-\lambda t} a \, dt. \end{aligned}$$

The proof of (2.35), using Fubini's theorem and the Riemann-Stieltjes integration by parts theorem, is left as an exercise. Using (2.35), we can rewrite

(2.31) as

$$(2.36) \qquad I = E\left[\int_0^\infty e^{-\lambda t}(\pi\, dB - wa\, dt - m\, dB)\right].$$

Now subtracting (2.29) from (2.36) we get

$$(2.37) \quad I - V = E\left\{\int_0^\infty e^{-\lambda t}\left[\pi\, dL + w(k - a)\, dt + pZ_t\, dt\right.\right.$$

$$\left.\left. + m(k\, dt - dU - dB)\right]\right\}.$$

With $Z_0 = 0$, we have $Z_t = (kt - U_t) - (B_t - L_t)$. Using this and integration by parts again, we find that

$$(2.38) \qquad \begin{aligned} \int_0^\infty e^{-\lambda t}(k\, dt - dU - dB) &= \int_0^\infty e^{-\lambda t}(dZ - dL) \\ &= \int_0^\infty e^{-\lambda t}(\lambda Z_t\, dt - dL). \end{aligned}$$

Substituting (2.38) into (2.37) and collecting similar terms, we have $I - V = \Delta$. $\qquad\qquad\qquad\qquad\qquad\qquad\qquad\qquad\qquad\qquad\qquad\qquad\qquad\qquad$ □

Obviously Δ represents the amount by which management's plan falls short, in expected present value terms, of the ideal profit level I. The definition (2.34) expresses this shortfall as the sum of three effects. First, the contribution margin δ is lost on each unit of potential sales foregone. Second, we continuously incur a cost of w dollars for each unit of capacity in excess of the average demand rate. Finally, for each unit of production held in inventory we continuously incur an out-of-pocket cost p plus an opportunity cost λm. We emphasize again that Δ measures the *degradation of system performance from a deterministic ideal*. Thus the minimum achievable Δ value may be viewed as the *cost of stochastic variability*.

Our first objective here is to develop a quantitative theory of storage system performance. As a natural outgrowth of that descriptive objective, we also seek to prescribe means by which management can minimize or at least reduce performance degradation, such as investment in excess capacity (a design decision) and maintenance of buffer stock (a matter of operating policy).

In concluding this section, let us briefly consider a cost structure in which $\lambda \downarrow 0$ but δ, w, and h remain constant. Further suppose that

$$(2.39) \qquad \frac{1}{t}E(L_t) \longrightarrow \alpha \quad \text{and} \quad E(Z_t) \longrightarrow \gamma \qquad \text{as } t \to \infty.$$

Obviously α represents a long-run average *lost sales rate*, whereas γ is the long-run average inventory level. Under mild additional assumptions, it is well known that $\lambda\Delta$ approaches the *long-run average cost rate*

(2.40) $\rho := \delta\alpha + w\mu + h\gamma$

as $\lambda \downarrow 0$. Thus minimizing Δ is approximately equivalent to minimizing ρ for small values of λ, and it is usually easier to calculate ρ than the discounted performance measure Δ.

2.6 Brownian storage models

Suppose that, in the setting of Section 2.3, we directly model the netflow process X as a (μ, σ) Brownian motion. The inventory process Z, lost potential output L, and lost potential input U are then defined by applying the two-sided reflection mapping to X exactly as before. In the obvious way, we call Z a *reflected Brownian motion* (RBM). It will be seen later that all the performance measures discussed in Section 2.5, and a number of other interesting quantities, can be calculated explicitly for RBM.

RBM is tractable, and therefore appealing as a storage model, but it is actually inconsistent with the model description given in Section 2.3: we have seen earlier that the sample paths of Brownian motion have infinite variation, and thus it cannot represent the difference between a potential input process and a potential output process. Nonetheless, a netflow process may be well approximated by Brownian motion under certain conditions. To understand these conditions, recall that Brownian motion is the *unique* stochastic process having stationary, independent increments and continuous sample paths; unbounded variation follows as a consequence of these primitive properties. Also note that the total variation of a netflow process over any given interval equals the *sum* of potential input and potential output over that interval. If such a netflow process is to be well approximated by Brownian motion, both potential input and potential output must be large for intervals of moderate length, but their difference (netflow itself) must be moderate in value. We may express this state of affairs by saying that we have a system of *balanced high-volume flows*.

Pulling together several themes, we conclude that Brownian motion may reasonably approximate the netflow process for a system of *stationary, continuous, balanced high-volume flow, where netflow increments during non-overlapping intervals are approximately independent*. Formal limit theorems that give this statement precise mathematical form, and thus serve to justify Brownian approximations, have been proved for various types of storage systems. The Brownian model discussed in this section will be studied extensively in future chapters, and readers should keep in mind its domain of applicability.

2.7 Problems and complements

Problem 2.1 Prove Proposition 2.3, thus verifying the one-sided reflection mapping's lack of memory.

Problem 2.2 Prove that $l := \psi(x)$ satisfies (2.9) to (2.11).

Figure 2.4 A tandem buffer system.

Problem 2.3 Consider the tandem buffer system pictured in Figure 2.4. Each buffer has infinite capacity, and we denote by $X_k(0)$ the initial inventory in buffer k. Extending in an obvious way the model of Section 2.1, we take as primitive three increasing, continuous processes $A_k = \{A_k(t), t \geq 0\}$ such that $A_k(0) = 0$ ($k = 1, 2, 3$). Interpret A_1 as input to the first buffer, A_2 as potential transfer between the two buffers, and A_3 as potential output from the second buffer. Define a (continuous) vector netflow process $X(t) := [X_1(t), X_2(t)]$ by setting

$$X_1(t) = X_1(0) + A_1(t) - A_2(t) \qquad \text{for } t \geq 0$$

and

$$X_2(t) = X_2(0) + A_2(t) - A_3(t) \qquad \text{for } t \geq 0.$$

Let $L_2(t)$ denote the amount of the potential transfer $A_2(t)$ that is lost over $[0, t]$ because of emptiness of the first buffer, and define $L_3(t)$ in the obvious analogous fashion. Let $Z_k(t)$ denote the content of buffer k at time t. Applying the analysis of Section 2.1 and Section 2.2 first to buffer 1 and then to buffer 2 in isolation, show that $L_2 = \psi(X_1)$, $Z_1 = \phi(X_1)$, $L_3 = \psi(X_2 - L_2)$, and $Z_2 = \phi(X_2 - L_2)$. Conclude that $L := (L_2, L_3)$ and $Z := (Z_1, Z_2)$ uniquely satisfy the following conditions: (a) L_2 and L_3 are increasing and continuous with $L_2(0) = L_3(0) = 0$; (b) $Z_1(t) = X_1(t) + L_2(t) \geq 0$ for all $t \geq 0$, and $Z_2(t) = X_2(t) - L_2(t) + L_3(t) \geq 0$ for all $t \geq 0$; and (c) L_2 and L_3 increase only when $Z_1 = 0$ and $Z_2 = 0$, respectively.

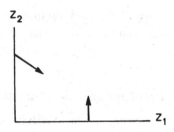

Figure 2.5 Directions of reflection for a tandem buffer system.

All of this describes the mapping by which (L, Z) is obtained from X. (It is again important that L and Z depend on primitive model elements only through the netflow process X.) Conditions (a) to (c) suggest the following interpretation or animation of that path-to-path transformation. An observer watches $X = (X_1, X_2)$ and may increase at will either component of a cumulative control process $L = (L_2, L_3)$. These actions determine $Z = (Z_1, Z_2)$ according to (b). The observer increases L_2 only as necessary to ensure that $Z_1 \geq 0$, so L_2 increases only when $Z_1 = 0$. Each such increase causes a positive displacement of Z_1 (or rather prevents a negative one) *and an equal negative displacement of* Z_2. Thus the effect of the observer's action at $Z_1 = 0$ is to drive Z in the diagonal direction pictured in Figure 2.5. On the other hand, L_3 is increased at the boundary $Z_2 = 0$ so as to ensure $Z_2 \geq 0$, producing only the vertical displacement pictured in Figure 2.5. Hereafter we shall say that (L, Z) is obtained by applying a *multidimensional reflection mapping* to X, the control region and *directions of reflection* being as illustrated in Figure 2.5. (As noted in the introduction, a term like "direction of control" would be more descriptive than "direction of reflection," but the latter term is fixed by historical use.) This problem is adapted from Harrison (1978).

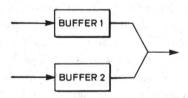

Figure 2.6 An assembly or blending operation.

Problem 2.4 A similar sort of multi-dimensional storage system is pictured in Figure 2.6. Here there are two input processes, each feeding its own infinite storage buffer. These inputs are then combined, exactly one unit of each input being required to produce one unit of system output. (The important point here is that inputs are combined in *fixed proportions*; the rest is just a matter of how units are defined.) This is the structure of an assembly operation, but again we treat the system flows as if they were continuous so that attention is effectively restricted to high-volume assembly systems. For another application, Figure 2.6 might be interpreted as a blending operation in which liquid or granulated ingredients are combined in fixed proportions to produce a similarly continuous output. To build a model, we again take as primitive initial inventory levels $X_1(0) \geq 0$ and $X_2(0) \geq 0$ plus three increasing, continuous processes $A_k = \{A_k(t), t \geq 0\}$ with $A_k(0) = 0$ ($k = 1, 2, 3$). Interpret A_1 and A_2 as input to buffer 1 and buffer 2, respectively, and A_3 as potential output. Potential output is lost if *either* buffer is empty, and we denote by $L(t)$ the cumulative potential output lost up to time t because of such emptiness. For purposes of determining L, the blending operation may be viewed as a simple storage system with initial inventory plus cumulative input given by

$$A^*(t) := [X_1(0) + A_1(t)] \wedge [X_2(0) + A_2(t)].$$

Let $Z_k(t)$ denote the inventory level in buffer k at time t, and define a (continuous) vector netflow process $X = [X_1(t), X_2(t)]$ by setting

$$X_1(t) := X_1(0) + A_1(t) - A_3(t) \qquad \text{for } t \geq 0$$

and

$$X_2(t) := X_2(0) + A_2(t) - A_3(t) \qquad \text{for } t \geq 0.$$

Applying the results of Section 2.1 and Section 2.2, write out explicit formulas for L and $Z := (Z_1, Z_2)$ in terms of X. (Again it is important that L and Z depend on primitive model elements only through the netflow process X.) Conclude that L and Z together uniquely satisfy the following conditions: (a) L is increasing and continuous with $L(0) = 0$; (b) $Z_1(t) = X_1(t) + L(t) \geq 0$ for all $t \geq 0$, and $Z_2(t) = X_2(t) + L(t) \geq 0$ for all $t \geq 0$; and (c) L increases only when $Z_1 = 0$ or $Z_2 = 0$.

The mapping that carries X into (L, Z) may be pictured as in Figure 2.7. The inventory process Z coincides with X up until X hits the boundary of the positive quadrant. At that point, L increases, causing *equal positive displacements in both Z_1 and Z_2* as necessary to keep $Z_1 \geq 0$ and $Z_2 \geq 0$.

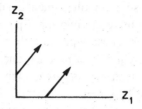

Figure 2.7 Directions of reflection for a blending operation.

Thus the effect of increases in L at the boundary is to drive Z in the diagonal direction shown in Figure 2.7, regardless of which boundary surface is struck. This problem is adapted from Harrison (1973).

Problem 2.5 Assuming for convenience that $x_0 = 0$, write out an explicit recursive expression for the times $T_1 < T_2 < \dots$ identified in the proof of Proposition 2.4. Show that if $T_n \uparrow T < \infty$, then x cannot be continuous at T; thus $T_n \to \infty$ as $n \to \infty$.

Problem 2.6 Consider again the tandem buffer system of Problem 2.3, assuming that buffers 1 and 2 now have *finite* capacities b_1 and b_2, respectively. In this case, potential input is lost when the first buffer is full, and potential transfer is lost when *either* the first buffer is empty *or* the second one is full. (We say that the transfer process is *starved* in the former case and *blocked* in the latter case.) In addition to the notation established in Problem 2.3, let $L_1(t)$ denote total potential input lost up to time t. Argue that $L := (L_1, L_2, L_3)$ and $Z := (Z_1, Z_2, Z_3)$ should jointly satisfy

(a) L_k is continuous and increasing with
             $L_k(0) = 0$ $(k = 1, 2, 3)$.

(b) $Z_1(t) := X_1(t) + L_2(t) - L_1(t) \in [0, b_1]$ for all $t \geq 0$,
             $Z_2(t) := X_2(t) + L_3(t) - L_2(t) \in [0, b_2]$ for all $t \geq 0$.

(c) L_1 increases only when $Z_1 = b_1$,
             L_2 increases only when $Z_1 = 0$ or $Z_2 = b_2$, and
             L_3 increases only when $Z_2 = 0$.

Explain the connection between (a) to (c) and Figure 2.8. Describe informally how one can use the results of Problems 2.3 and 2.4 to prove existence and uniqueness of a pair (L, Z) satisfying (a) to (c). This problem is adapted from Wenocur (1982).

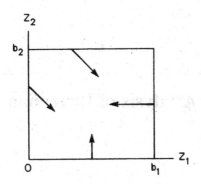

Figure 2.8 Directions of reflection for finite buffers in tandem.

Problem 2.7 Show that the linearized random walk model of demand, described in Section 2.5, satisfies (2.27) and (2.28).

Problem 2.8 It was assumed in Section 2.5 that overtime production was impossible. Suppose instead that unlimited amounts of overtime production are available at a premium wage rate $w^* > w$, regardless of what workforce level may be chosen at the beginning. To keep things simple, assume that overtime production is instantaneous. (One may also think in terms of buying finished goods at a premium price from some alternate supplier and then using those goods to satisfy demand.) Finally, assume that $\pi - w^* - m > 0$, so it is always better to use overtime production than to forgo potential sales. The basic structure of this system is identical to that discussed in Section 2.5, but now L_t is interpreted as cumulative overtime production up to time t. Show that maximizing the expected present value of total profit is equivalent to minimizing Δ, where Δ is given by formula (2.34) with w^* in place of δ.

Problem 2.9 Prove the three equalities of (2.35), using Fubini's theorem (Section A.5) and the Riemann–Stieltjes integration by parts theorem (Section B.3).

3

Further Analysis of Brownian Motion

The treatment of Brownian motion in Chapter 1 was restricted to the case $X_0 = 0$. As we move on to more complex calculations, it will be convenient to view the starting state as a variable parameter. This is accomplished by introducing a family of probability measures on path space, with each member of the family corresponding to a different starting state.

3.1 Introduction

Throughout this chapter we take $\Omega = C$, denote by $\mathcal{C}$ the Borel σ-algebra on Ω (see Section A.2), and let X be the coordinate process on Ω as in Section A.3. Let μ and $\sigma > 0$ be fixed constants. For each $x \in \mathbb{R}$ there is a unique probability measure P_x on $(\Omega, \mathcal{C})$ such that

(3.1) $\qquad X$ is a (μ, σ) Brownian motion on $(\Omega, \mathcal{C}, P_x)$

and

(3.2) $\qquad P_x \{\omega \in \Omega : X_0(\omega) = x\} = 1.$

This follows from Wiener's Theorem 1.1. We paraphrase (3.1) and (3.2) by saying that X is a (μ, σ) Brownian motion *with starting state* x under P_x. Heuristically, one may think of $P_x(A)$ as the conditional probability of event A given that $X_0 = x$.

Fixing a starting state $x \in \mathbb{R}$, we adopt the canonical setup $(\Omega, \mathcal{F}, \mathbb{F}, P_x)$ described in Section A.3, which means that $\mathcal{F}$ is the P_x-completion of $\mathcal{C}$ and $\mathbb{F}$ is the filtration generated by X. Also, let E_x be the expectation operator associated with P_x. That is,

$$E_x(Z) := \int_\Omega Z(\omega) P_x(d\omega)$$

for all measurable functions (random variables) $Z : \Omega \to \mathbb{R}$ such that the integral on the right exists.

With this setup, the strong Markov property of X can be recast in the following form. Let T be an arbitrary stopping time and set

(3.3) $$X_t^* := X_{T+t}, \quad t \geq 0, \quad \text{on } \{T < \infty\}.$$

(This is *not*, however, the definition of X^* that was used in Section 1.3.) More precisely, (3.3) means that

$$X_t^*(\omega) := X_{T(\omega)+t}(\omega), \quad t \geq 0,$$

for ω such that $T(\omega) < \infty$. The process X^* need not be defined at all on $\{\omega : T(\omega) = \infty\}$. Now let F be a measurable mapping $(C, \mathcal{C}) \rightarrow (\mathbb{R}, \mathcal{B})$ such that $E_x\{|F(X)|\} < \infty$ for all $x \in \mathbb{R}$ and define

(3.4) $$f(x) := E_x[F(X)], \quad x \in \mathbb{R}.$$

From our original articulation in Theorem 1.7 of the strong Markov property it follows that

(3.5) $$E_x[F(X^*)|\mathcal{F}_T] = f(X_0^*) := f(X_T) \quad \text{on } \{T < \infty\}.$$

This relationship plays a major role in Section 3.5 and Section 3.6, where we calculate expected discounted costs for Brownian motion with absorbing barriers.

3.2 The backward and forward equations

Recall that $X_{t+s} - X_t \sim N(\mu s, \sigma^2 s)$ under P_x for each $x \in \mathbb{R}$. Thus the transition density

$$p(t, x, y)\, dy := P_x\{X_t \in dy\} \quad \text{for } t \geq 0 \text{ and } x, y \in \mathbb{R}$$

is given by

(3.6) $$p(t, x, y) = (\sigma^2 t)^{-1/2} \phi\left(\frac{y - x - \mu t}{\sigma t^{1/2}}\right)$$

where $\phi(z) := (2\pi)^{-1/2} \exp(-z^2/2)$ is the standard normal density function. Direct calculation shows (see Problem 3.1) that p satisfies

(3.7) $$\frac{\partial}{\partial t} p(t, x, y) = \left(\frac{1}{2}\sigma^2 \frac{\partial^2}{\partial x^2} + \mu \frac{\partial}{\partial x}\right) p(t, x, y)$$

with initial condition

(3.8) $$p(0, x, y) = \delta(x - y).$$

Here $\delta(\cdot)$ is the Dirac delta function; (3.8) is defined to mean that

$$\int_{\mathbb{R}} h(y)p(t,x,y)\,dy \longrightarrow h(x) \qquad \text{as } t \downarrow 0$$

for all bounded, continuous $h : \mathbb{R} \to \mathbb{R}$. This differential equation for the transition density p will not play much of a role in our future analysis of Brownian motion. It has, however, played a major role in the historical development of the subject, and a little more commentary is appropriate. In probability theory, (3.7) is called *Kolmogorov's backward equation* for the Markov process X. Computing directly from (3.6), readers may verify the corresponding *forward equation*

$$(3.9) \qquad \frac{\partial}{\partial t} p(t,x,y) = \left(\frac{1}{2}\sigma^2 \frac{\partial^2}{\partial y^2} - \mu \frac{\partial}{\partial y} \right) p(t,x,y).$$

Note that we differentiate with respect to the *backward variable* (initial state) x on the right side of (3.7), whereas (3.9) involves differentiation with respect to the *forward variable* (final state) y. In the special case where $\mu = 0$, equation (3.9) reduces to the celebrated *heat equation* (or diffusion equation) of mathematical physics. Because of this connection with the mathematics of physical diffusion, Brownian motion and certain of its close relatives are called *diffusion processes*.

3.3 Hitting time problems

Hereafter let $T(y)$ denote the first time $t \geq 0$ at which $X_t = y$, with $T(y) = \infty$ if no such t exists. Fixing $b > 0$, we restrict attention to starting states $X_0 = x \in [0,b]$ and let $T := T(0) \wedge T(b)$ as in Figure 3.1. The objective in this section is to calculate the Laplace transform $E_x[\exp(-\lambda T)]$ and various other related quantities. In particular, we shall compute the Laplace transforms of $T(0)$ and $T(b)$ and the probability that level b is hit before level zero. It will be seen later that the formulas derived in this section are indispensable for computing expected discounted costs.

Proposition 3.1 $E_x(T) < \infty$, $0 \leq x \leq b$.

Proof First consider the case $\mu > 0$. Let $M_t := X_t - \mu t$ for $t \geq 0$. A slight modification of the argument given in Section 1.4 shows that M is a martingale on $(\Omega, \mathbb{F}, P_x)$ and thus the martingale stopping theorem (see Section A.4) says that $E_x[M(T \wedge t)] = E_x(M_0)$. That is,

$$(3.10) \qquad E_x[X(T \wedge t)] - \mu E_x(T \wedge t) = x.$$

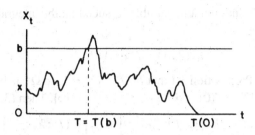

Figure 3.1 First exit time T from $[0, b]$.

Of course, $X(T \wedge t) \leq b$, so $E_x(T \wedge t) \leq (b - x)/\mu$ by (3.10). Because this holds for any $t > 0$, we have $E_x(T) \leq (b - x)/\mu < \infty$. The case $\mu < 0$ is handled symmetrically. Finally, if $\mu = 0$, it follows from Proposition 1.8 that $\{X_t^2 - \sigma^2 t, \; t \geq 0\}$ is a martingale on $(\Omega, \mathbb{F}, P_x)$. Then the martingale stopping theorem gives

$$(3.11) \qquad E_x \left[X^2(T \wedge t) - \sigma^2(T \wedge t) \right] = E_x(X_0^2) = x^2$$

for any $t > 0$. But $X^2(T \wedge t) \leq b^2$ and therefore (3.11) gives $E_x(T \wedge t) \leq (b^2 - x^2)/\sigma^2$ for all $t > 0$, thus implying $E_x(T) \leq (b^2 - x^2)/\sigma^2 < \infty$. □

Recall that in Section 1.4 the Wald martingale V_β with dummy variable $\beta \in \mathbb{R}$ was defined via

$$(3.12) \qquad V(t) := \exp[\beta X_t - q(\beta)t], \qquad t \geq 0,$$

where

$$(3.13) \qquad q(\beta) := \mu\beta + \frac{\sigma^2 \beta^2}{2}.$$

(Again the argument given in Section 1.4 must be modified slightly to show that V_β is a martingale in our current setting, but readers should have no trouble supplying the details.) Hereafter we restrict attention to β values such that $q(\beta) \geq 0$. Then $\{V_\beta(T \wedge t), \; t \geq 0\}$ is a *bounded* family of random variables, and Corollary A.4 of the martingale stopping theorem gives

$$(3.14) \qquad E_x \left[V_\beta(T) \right] = E_x \left[V_\beta(0) \right] = e^{\beta x}, \qquad 0 \leq x \leq b.$$

To further develop this identity, we define the *partial expectation*

$$(3.15) \qquad E_x(Z; A) := \int_A Z \, dP_x.$$

for events $A \in \mathcal{F}$ and random variables Z such that the integral on the right exists. Note that

(3.16) $$E_x(Z; A) = E_x(Z|A)P_x(A).$$

It follows from Proposition 3.1 that $P_x\{T < \infty\} = 1$, so Ω can be partitioned into the events $\{T = T(0) < \infty\}$ and $\{T = T(b) < \infty\}$. Then (3.14) gives

(3.17)
$$e^{\beta x} = E_x\left[V_\beta(T); X_T = 0\right] + E_x\left[V_\beta(T); X_T = b\right]$$
$$= E_x\left[e^{-q(\beta)T}; X_T = 0\right] + E_x\left[e^{\beta b - q(\beta)T}; X_T = b\right].$$

To repeat, (3.17) holds for all $x \in [0, b]$ and all β such that $q(\beta) \geq 0$. Now for $\lambda > 0$ and $0 \leq x \leq b$ let us define

(3.18) $$\psi_1(x|\lambda) := E_x(e^{-\lambda T}; X_T = 0)$$

and

(3.19) $$\psi_2(x|\lambda) := E_x(e^{-\lambda T}; X_T = b).$$

Note that (3.17) can be re-expressed as

(3.20) $$e^{\beta x} = \psi_1(x|q(\beta)) + e^{\beta b}\psi_2(x|q(\beta)).$$

Equation (3.20) will be used shortly to compute ψ_1 and ψ_2. The Laplace transform of T (with dummy variable λ) is given by

(3.21) $$E_x(e^{-\lambda T}) = \psi_1(x|\lambda) + \psi_2(x|\lambda), \qquad 0 \leq x \leq b,$$

but one needs to know the terms ψ_1 and ψ_2 individually to compute expected discounted costs (see Sections 3.4 to 3.6). From the definition (3.13) of $q(\cdot)$ we see that the *two* values of β that yield $q(\beta) = \lambda > 0$ are $\beta = \alpha_2(\lambda)$ and $\beta = -\alpha_1(\lambda)$, where

(3.22) $$\alpha_1(\lambda) := (1/\sigma^2)\left[(\mu^2 + 2\sigma^2\lambda)^{1/2} + \mu\right] > 0$$

and

(3.23) $$\alpha_2(\lambda) := (1/\sigma^2)\left[(\mu^2 + 2\sigma^2\lambda)^{1/2} - \mu\right] > 0.$$

These two roots are pictured in Figure 3.2 for a case where $\mu > 0$. (Note that $q'(0) = \mu$.) Substitution of $\beta = -\alpha_1(\lambda)$ and $\beta = \alpha_2(\lambda)$ into (3.20) gives

(3.24) $$e^{-\alpha_1(\lambda)x} = \psi_1(x|\lambda) + e^{-\alpha_1(\lambda)b}\psi_2(x|\lambda)$$

and

(3.25) $$e^{\alpha_2(\lambda)x} = \psi_1(x|\lambda) + e^{\alpha_2(\lambda)b}\psi_2(x|\lambda).$$

Hereafter we suppress the dependence of ψ_1 and ψ_2 on λ. Solving (3.24) and (3.25) simultaneously gives the following.

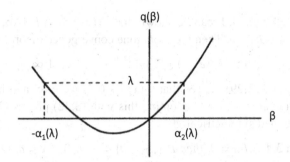

Figure 3.2 The two roots of $q(\beta) = \lambda > 0$.

Proposition 3.2 *Let $\lambda > 0$ be fixed. For $0 \leq x \leq b$,*

$$(3.26) \qquad \psi_2(x) = \frac{\theta_2(x) - \theta_1(x)\,\theta_2(0)}{1 - \theta_1(b)\,\theta_2(0)}$$

and

$$(3.27) \qquad \psi_1(x) = \frac{\theta_1(x) - \theta_2(x)\,\theta_1(b)}{1 - \theta_1(b)\,\theta_2(0)}$$

where

$$(3.28) \qquad \theta_1(x) := \exp\{-\alpha_1(\lambda)x\}$$

and

$$(3.29) \qquad \theta_2(x) := \exp\{-\alpha_2(\lambda)(b - x)\}.$$

From the basic formulas (3.26) and (3.27) a variety of useful corollaries can be extracted. In the development to follow, let us agree to write

$$E_x[e^{-\lambda T(y)}] := E_x\left[e^{-\lambda T(y)}; T(y) < \infty\right]$$

for $\lambda > 0$ and $x, y \in \mathbb{R}$.

Proposition 3.3 *Let θ_1 and θ_2 be defined by (3.28) and (3.29) respectively. Then*

$$(3.30) \qquad E_x[e^{-\lambda T(0)}] = \theta_1(x), \qquad 0 \leq x \leq b$$

and

$$(3.31) \qquad E_x[e^{-\lambda T(b)}] = \theta_2(x), \qquad 0 \leq x \leq b.$$

Proof Let x be fixed. It can be shown that $T(b) \uparrow \infty$ as $b \uparrow \infty$, implying that $T \uparrow T(0)$ as $b \uparrow \infty$. Then the monotone convergence theorem gives

$$(3.32) \qquad \psi_1(x) := E_x(e^{-\lambda T}) \downarrow E_x[e^{-\lambda T(0)}] \qquad \text{as } b \uparrow \infty.$$

From (3.27) and (3.29) we see that $\theta_2(x) \to 0$ as $b \to \infty$, and hence that $\psi_1(x) \to \theta_1(x)$ as $b \to \infty$. Combining this with (3.32) proves (3.30), and (3.31) is obtained symmetrically. $\qquad\qquad \square$

Proposition 3.4 *If $\mu = 0$, then $P_x\{X_T = b\} = x/b$, $0 \leq x \leq b$. Otherwise,*

$$(3.33) \qquad P_x\{X_T = b\} = \frac{1 - \xi(x)}{1 - \xi(b)}, \qquad 0 \leq x \leq b,$$

where

$$(3.34) \qquad \xi(z) := \exp\left(\frac{-2\mu z}{\sigma^2}\right).$$

Corollary 3.5 *If $\mu \leq 0$, then $P_x\{T(0) < \infty\} = 1$ for all $x \geq 0$. If $\mu > 0$, then $P_x\{T(0) < \infty\} = \xi(x)$ for $x > 0$.*

Proof The monotone convergence theorem gives

$$\psi_2(x) := E_x(e^{-\lambda T}; X_T = b) \uparrow P_x\{X_T = b\} \qquad \text{as } \lambda \downarrow 0.$$

Proposition 3.4 follows immediately from this and the formula (3.26) for ψ_2. The corollary is then obtained by letting $\lambda \downarrow 0$ in the formula developed earlier for $E_x[\exp\{-\lambda T(0)\}]$. $\qquad\qquad \square$

In the interest of efficiency, we have deduced Proposition 3.3, Proposition 3.4, and Corollary 3.5 from the master transform relation that is Proposition 3.2. However, one can obtain Proposition 3.3, Proposition 3.4, and Corollary 3.5 directly by using the Wald martingale V_β and the martingale stopping theorem.

For future reference, it will be useful to observe that each of the transforms computed in this section, viewed as a function of the starting state x, satisfies a characteristic second-order differential equation subject to particular boundary conditions. Specifically, if we define the differential operator

$$(3.35) \qquad \Gamma := \frac{1}{2}\sigma^2 \frac{d^2}{dx^2} + \mu \frac{d}{dx}$$

then it can be verified from the explicit formulas displayed above that (for $\lambda > 0$)

(3.36) $\qquad \lambda\theta_1 - \Gamma\theta_1 = \lambda\theta_2 - \Gamma\theta_2 = 0 \qquad$ in $\mathbb{R}$,

(3.37) $\qquad \theta_1(0) = \theta_2(b) = 1, \quad$ and $\quad \theta_1(\infty) = \theta_2(-\infty) = 0.$

Consequently,

(3.38) $\qquad \lambda\psi_1 - \Gamma\psi_1 = \lambda\psi_2 - \Gamma\psi_2 = 0 \qquad$ in $(0, b)$,

(3.39) $\qquad \psi_1(0) = \psi_2(b) = 1, \quad$ and $\quad \psi_1(b) = \psi_2(0) = 0.$

We have solved these simple ordinary differential equations (ODEs) by probabilistic methods, and more particularly by manipulation of the Wald martingale for Brownian motion. The relationship between Brownian motion and various differential equations will be developed further in the problems at the end of this chapter and in later chapters.

3.4 Expected discounted costs

Let $\lambda > 0$ be fixed, and let u be a continuous function on $\mathbb{R}$ such that $|u|$ is bounded by a polynomial. It follows that

(3.40) $\qquad \displaystyle\int_0^\infty e^{-\lambda t} E_x \{|u(X_t)|\} \, dt < \infty \qquad$ for all $x \in \mathbb{R}.$

Now define

(3.41) $\quad f(x) := E_x \left[\displaystyle\int_0^\infty e^{-\lambda t} u(X_t) \, dt \right] = \int_0^\infty e^{-\lambda t} E_x \left[u(X_t) \right] dt, \qquad x \in \mathbb{R}.$

The second equality in (3.41) follows from Fubini's theorem (Section A.5). We interpret $u(y)$ as the rate at which costs are incurred when $X(t) = y$ and λ as the *interest rate* appropriate for discounting (see Section 2.5); thus $f(x)$ represents the *expected discounted cost* incurred over an infinite horizon, starting from level x. For certain specific functions u, one can explicitly calculate $E[u(X_t)]$ for general t and then perform the integration in (3.41). For example, if $u(x) = x$, we have $E_x[u(X_t)] = x + \mu t$, and a simple integration gives the following.

Proposition 3.6 *If $u(x) = x$, then $f(x) = x/\lambda + \mu/\lambda^2$.*

An equally simple formula for $f(x)$ can be obtained in this way if the cost function $u(\cdot)$ is quadratic (see Problem 3.4). The next proposition gives a general formula for $f(x)$ as the integral of $u(\cdot)$ against a certain kernel.

Proposition 3.7 $f(x) = \int_{-\infty}^{\infty} u(y)\pi(x,y)\,dy,\ x \in \mathbb{R},\ where$

(3.42) $\pi(x,y) = (\mu^2 + 2\lambda\sigma^2)^{-1/2}\theta(x,y)$

and

(3.43) $\theta(x,y) = \begin{cases} \exp\left[-\alpha_1(\lambda)(x-y)\right] & \text{if } x \geq y \\ \exp\left[-\alpha_2(\lambda)(y-x)\right] & \text{if } y \geq x. \end{cases}$

Because this general formula is not used later, we shall merely sketch its proof in Problems 3.5 to 3.8, where an interpretation for $\pi(x,y)$ is also given in terms of Brownian local time. To display the differential equation satisfied by the kernel π, let us fix y and agree to write $\pi(x) := \pi(x,y)$. Readers may verify that π is continuous, is twice continuously differentiable except at $x = y$, and satisfies

(3.44) $\lambda\pi(x) - \Gamma\pi(x) = 0$ except at $x = y$

and

(3.45) $\dfrac{1}{2}\sigma^2\Delta\pi'(y) = -1$

where Γ is the differential operator defined by (3.35) and $\Delta\pi'(y)$ is the jump in the derivative of $\pi(\cdot)$ going from left to right through $x = y$. It is left as an exercise to show, using Proposition 3.7, (3.44), and (3.45), that

(3.46) $\lambda f(x) - \Gamma f(x) = u(x)$ for all $x \in \mathbb{R}$.

This relationship between the cost function u and the expected discounted cost f will be studied further in Problem 4.3 and Problem 4.4.

3.5 One absorbing barrier

Now we restrict attention to positive starting states x, set $T := T(0)$ throughout this section, and define $Y_t := X(T \wedge t)$, $t \geq 0$. Thus Y is a (μ, σ) Brownian motion with starting state x and *absorption at the origin* under P_x. The first goal of this section is to compute (for $x, y \geq 0$)

(3.47) $G(t,x,y) := P_x\{Y_t > y\} = P_x\{X_t > y;\ T(0) > t\}.$

Recall that in Section 1.7 we derived the joint distribution of X_t and M_t in the case $X_0 = 0$, where $M_t := \sup\{X_s, \ 0 \le s \le t\}$. From this we can deduce that

$$(3.48) \qquad G(t, x, y) = \Phi\left(\frac{-y + x + \mu t}{\sigma t^{1/2}}\right) - \exp\left(\frac{-2\mu x}{\sigma^2}\right)\Phi\left(\frac{-y - x + \mu t}{\sigma t^{1/2}}\right).$$

In Problem 4.13 this formula will be verified by independent means, using the fact that (3.48) satisfies

$$(3.49) \qquad \frac{\partial}{\partial t}G(t, x, y) = \left(\frac{1}{2}\sigma^2\frac{\partial^2}{\partial x^2} + \mu\frac{\partial}{\partial x}\right)G(t, x, y) \quad \text{main equation,}$$

$$(3.50) \qquad G(t, 0, y) = 0 \qquad\qquad\qquad\qquad\qquad \text{boundary condition,}$$

$$(3.51) \qquad G(0, x, y) = 1_{(x>y)} \qquad\qquad\qquad\qquad \text{initial condition.}$$

Let $u : \mathbb{R} \to \mathbb{R}$ be a continuous cost function satisfying (3.40) as before. Building on earlier calculations, we now compute

$$(3.52) \qquad\qquad g(x) := E_x\left[\int_0^T e^{-\lambda t}u(X_t)\,dt\right], \qquad x \in \mathbb{R}.$$

This represents the *expected discounted cost incurred up to the time of absorption*. The first thing to note is that

$$(3.53) \qquad\qquad g(x) = f(x) - E_x\left[\int_T^\infty e^{-\lambda t}u(X_t)\,dt;\ T < \infty\right]$$

where $f(x)$ is the *infinite-horizon* expected discounted cost calculated in Section 3.4. On $\{T < \infty\}$, let $X_t^* := X_{T+t}$ and note that

$$(3.54) \qquad\qquad \int_T^\infty e^{-\lambda t}u(X_t)\,dt = e^{-\lambda T}\int_0^\infty e^{-\lambda t}u(X_t^*)\,dt.$$

To compute the second term on the right side of (3.53), we use the strong Markov property (3.5) with the particular functional

$$(3.55) \qquad\qquad F(X) := \int_0^\infty e^{-\lambda t}u(X_t)\,dt.$$

Specifically, (3.54), (3.55), and (3.5) give

$$E_x\left[\int_T^\infty e^{-\lambda t}u(X_t)\,dt;\ T<\infty\right]$$

$$= E_x\left[e^{-\lambda T}\int_0^\infty e^{-\lambda t}u(X_t^*)\,dt;\ T<\infty\right]$$

$$:= \int_{\{T<\infty\}}\left[e^{-\lambda T}\int_0^\infty e^{-\lambda t}u(X_t^*)\,dt\right]dP_x$$

$$= \int_{\{T<\infty\}}E_x\left[e^{-\lambda T}\int_0^\infty e^{-\lambda t}u(X_t^*)\,dt\,\Big|\,\mathcal{F}_T\right]dP_x$$

(3.56)

$$= \int_{\{T<\infty\}}e^{-\lambda T}E_x\left[\int_0^\infty e^{-\lambda t}u(X_t^*)\,dt\,\Big|\,\mathcal{F}_T\right]dP_x$$

$$= \int_{\{T<\infty\}}e^{-\lambda T}E_x[F(X^*)|\mathcal{F}_T]\,dP_x$$

$$= \int_{\{T<\infty\}}e^{-\lambda T}f(X_T)\,dP_x$$

$$= f(0)\int_{\{T<\infty\}}e^{-\lambda T}\,dP_x = f(0)e^{-\alpha_1(\lambda)x}.$$

The last equality in (3.56) uses Proposition 3.3. We summarize all of this as follows.

Proposition 3.8　　$g(x) = f(x) - f(0)e^{-\alpha_1(\lambda)x}$, $x \geq 0$.

From (3.28), (3.36), (3.37), and (3.46) it follows that g satisfies

(3.57)　　　　　　　　　　$\lambda g - \Gamma g = u$　　　on $(0,\infty)$

and

(3.58)　　　　　　　　　　$g(0) = 0$.

Our solution of the inhomogeneous equation (3.57) with boundary condition (3.58) has a form familiar to students of differential equations. It is built from a *particular solution* f of the main equation (3.57) plus a function $\theta_1(x) = \exp\{-\alpha_1(\lambda)x\}$ satisfying the *homogeneous* equation $\lambda\theta_1 - \Gamma\theta_1 = 0$ with boundary condition $\theta_1(0) = 1$. (There is also the question of boundary conditions at infinity, but we need not go into this at present.) In the next section, a probabilistic solution will be derived for the analogous problem on a finite interval.

3.6 Two absorbing barriers

Fixing $b > 0$ as in Section 3.3, let us again set $T := T(0) \wedge T(b)$ and restrict attention to starting states $x \in [0, b]$. Under P_x the process $\{X(T \wedge t), t \geq 0\}$ is a (μ, σ) Brownian motion with starting state x and two absorbing barriers. The time-dependent distribution $P_x\{X(T \wedge t) \leq y\}$ is known only as an infinite sum, but again one can derive a fairly simple formula for the expected discounted cost incurred before absorption. Fix $\lambda > 0$, let $u : \mathbb{R} \to \mathbb{R}$ be a continuous cost function satisfying (3.40) as before, and define

$$
\begin{aligned}
h(x) &:= E_x\left[\int_0^T e^{-\lambda t} u(X_t)\, dt\right] \\
&= E_x\left[\int_0^\infty e^{-\lambda t} u(X_t)\, dt\right] - E_x\left[\int_T^\infty e^{-\lambda t} u(X_t)\, dt\right].
\end{aligned}
$$

(3.59)

The first term on the right side of (3.59) is the quantity $f(x)$ calculated in Section 3.4, whereas the second term can be expressed as

$$
(3.60) \quad E_x\left[\int_T^\infty e^{-\lambda t} u(X_t)\, dt;\ X_T = 0\right] + E_x\left[\int_T^\infty e^{-\lambda t} u(X_t)\, dt;\ X_T = b\right].
$$

Proceeding exactly as in (3.56), we define $X_t^* = X_{T+t}$ and use the strong Markov property (3.5) to conclude that

$$
\begin{aligned}
E_x\Bigg[\int_T^\infty & e^{-\lambda t} u(X_t)\, dt;\ X_T = 0\Bigg] \\
&= E_x\left[e^{-\lambda T}\int_0^\infty e^{-\lambda t} u(X_t^*)\, dt;\ X_T = 0\right] \\
&:= \int_{\{X(T)=0\}}\left[e^{-\lambda T}\int_0^\infty e^{-\lambda t} u(X_t^*)\, dt\right] dP_x \\
&= \int_{\{X(T)=0\}} E_x\left[e^{-\lambda T}\int_0^\infty e^{-\lambda t} u(X_t^*)\, dt\ \Big|\ \mathcal{F}_T\right] dP_x \\
&= \int_{\{X(T)=0\}} e^{-\lambda T} f(0)\, dP_x := f(0)\psi_1(x).
\end{aligned}
$$

In the same way, the second term in (3.60) reduces to $f(b)\psi_2(x)$, and we arrive at the following.

Proposition 3.9 *The expected discounted cost $h(x)$ defined in (3.59) is given by $h(x) = f(x) - f(0)\psi_1(x) - f(b)\psi_2(x)$, $0 \leq x \leq b$, where $\psi_1(\cdot)$ and $\psi_2(\cdot)$ are given by Proposition 3.2.*

From (3.38), (3.39), and (3.46) it follows that h satisfies

(3.61) $\lambda h - \Gamma h = u$ on $(0, b)$

and

(3.62) $h(0) = h(b) = 0.$

3.7 More on reflected Brownian motion

Restricting attention again to positive starting states x, let us form processes L and Z by applying to X the one-sided reflection mapping of Section 2.2. That is, let

$$L_t := \sup_{0 \le s \le t} X_s^- \quad \text{and} \quad Z_t := X_t + L_t \quad \text{for } t \ge 0,$$

implying that Z is a continuous process with $Z_0 = x$ under P_x. In Section 1.10 we calculated $P_0\{Z_t \le y\}$ using a time reversal argument. Using the joint distribution computed in Section 1.7, one can generalize this to show that

(3.63)
$$\begin{aligned}
Q(t, x, y) &:= P_x\{Z_t > y\} \\
&= \Phi\left(\frac{-y + x + \mu t}{\sigma t^{1/2}}\right) + e^{2\mu y/\sigma^2} \Phi\left(\frac{-y - x - \mu t}{\sigma t^{1/2}}\right)
\end{aligned}$$

for $x, y, t \ge 0$. As with the expression for $G(t, x, y)$ derived in Section 3.5, we shall later verify (3.63) by independent means (see Problem 6.12). In preparation, readers are asked in Problem 3.11 to verify that

(3.64) $\dfrac{\partial}{\partial t} Q(t, x, y) = \left(\dfrac{1}{2}\sigma^2 \dfrac{\partial^2}{\partial x^2} + \mu \dfrac{\partial}{\partial x}\right) Q(t, x, y)$ main equation,

(3.65) $\dfrac{\partial}{\partial x} Q(t, 0, y) = 0$ boundary condition,

(3.66) $Q(0, x, y) = 1_{(x > y)}$ initial condition.

The mysterious thing here is the boundary condition (3.65) whose explanation must await the development of stochastic calculus.

3.8 Problems and complements

Problem 3.1   Verify that the transition density $p(t, x, y)$ given by (3.6) satisfies the backward equation (3.7) and forward equation (3.9). Use the fact that $\phi'(z) = -z\phi(z)$.

Problem 3.2   Let $l(t, y)$ be the local time at level y of the (μ, σ) Brownian motion X as in Section 1.2, and let $u : \mathbb{R} \to \mathbb{R}$ be bounded and continuous. Take E_x of both sides in (1.7) and use Fubini's theorem to conclude that

$$\int_0^t E_x [u(X_s)] \, ds = \int_{\mathbb{R}} u(y) E_x [l(t, y)] \, dy.$$

But $E_x[u(X_s)] = \int_{\mathbb{R}} u(y) p(s, x, y) \, dy$ by the definition of the transition density p. Conclude that

(3.67) $$p(t, x, y) = \frac{\partial}{\partial t} E_x [l(t, y)].$$

Problem 3.3   Verify that the functions θ_1, θ_2, ψ_1, and ψ_2 given by (3.26) to (3.29) satisfy (3.36) to (3.39). In Problem 4.2 these differential equations and boundary conditions will be used to verify some of the calculations done in Section 3.3.

Problem 3.4   Working directly from (3.41), show that $\lambda^3 f(x) = \lambda^2 x^2 + 2x\mu\lambda + 2\mu^2 + \sigma^2\lambda$ if $u(y) = y^2$. Verify that the proposed solution for f satisfies $\lambda f - \Gamma f = u$ on $\mathbb{R}$.

Problem 3.5   Let $u : \mathbb{R} \to \mathbb{R}$ be a continuous function satisfying (3.40), fix $\lambda > 0$, and define the expected discounted cost function $f(x)$ via (3.41). Working directly from (3.41), use Fubini's theorem to show that $f(x) = \int_{\mathbb{R}} u(y)\pi(x, y) \, dy$, where

(3.68) $$\pi(x, y) := \int_0^\infty e^{-\lambda t} p(t, x, y) \, dt.$$

Observing that $p(t, y, y) = \phi(-\mu t^{1/2}/\sigma)/\sigma t^{1/2}$, show that

$$\pi(y, y) = (\mu^2 + 2\lambda\sigma^2)^{-1/2} \qquad \text{for all } y \in \mathbb{R}.$$

Problem 3.6 (*Continuation*)   Let $l(t, y)$ be the local time of X at level y as in Problem 3.2. Substitute (3.67) into (3.68), then use Fubini's theorem and integration by parts to show that

$$\pi(x, y) = E_x \left[\int_0^\infty e^{-\lambda t} l(dt, y) \right],$$

with the integral on the right defined path by path in the Riemann–Stieltjes sense.

Problem 3.7 (*Continuation*) Let $T := T(y)$, let X^* be defined in terms of T on $\{T < \infty\}$ by (3.3), and let $l^*(t, y)$ be the local time at level y for the process X^*. Recall from Section 1.2 that $l(t, y) = 0$ for $0 \le t \le T$. Show that

$$\int_0^\infty e^{-\lambda t}\, l(dt, y) = e^{-\lambda T} \int_0^\infty e^{-\lambda t}\, l^*(dt, y) \qquad \text{on } \{T < \infty\}.$$

Problem 3.8 (*Continuation*) Use the strong Markov property (3.5) and the results of Problems 3.6 and 3.7 to show that

$$(3.69) \qquad \qquad \pi(x, y) = E_x[e^{-\lambda T(y)}]\pi(y, y).$$

Let $\theta(x, y)$ be defined by (3.43). From Proposition 3.3 we see that $\theta(x, y) = E_x\{\exp[-\lambda T(y)]\}$. Combining this with (3.69) and the result of Problem 3.5, we obtain the general formula of Proposition 3.7 for the expected discounted cost $f(x)$.

Problem 3.9 With $\pi(\cdot, \cdot)$ defined by (3.43), verify that π satisfies (3.44) and (3.45). Using this, verify that f satisfies the differential equation (3.46).

Problem 3.10 Verify that our solution (3.48) for $G(t, x, y)$ satisfies the partial differential equation (3.49) with boundary condition (3.50) and initial condition (3.51). It is helpful to note that each term on the right side of (3.48) satisfies the main equation (3.49) separately.

Problem 3.11 Verify that our solution (3.63) for $Q(t, x, y)$ satisfies the partial differential equation (3.64) with boundary condition (3.65) and initial condition (3.66). Again it is helpful to note that each term satisfies the main equation separately.

Problem 3.12 Extend the proof of Proposition 3.1 to show that $E_x(T) = x(b - x)/\sigma^2$ if $\mu = 0$. For nonzero drift, show that

$$\mu E_x(T) = b\left[\frac{1 - \xi(x)}{1 - \xi(b)}\right] - x,$$

using Proposition 3.4 and the fact that $X_t - \mu t$ is a martingale on $(\Omega, \mathbb{F}, P_x)$.

Problem 3.13 Fix $b > 0$, assume $0 \le x \le b$, and let $T = T(0) \wedge T(b)$ as in Section 3.3. Consider a process Z that coincides with X over $[0, T)$ but then jumps instantaneously to q or Q, where $0 < q < Q < b$, depending on whether $X_T = 0$ or $X_T = b$. Thereafter Z repeats this behavior in a

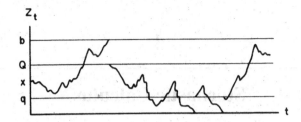

Figure 3.3 Brownian motion with jump boundaries.

regenerative fashion as shown in Figure 3.3. Our purpose is to compute the expected discounted cost function

$$k(x) := E_x \left[\int_0^\infty e^{-\lambda t} u(Z_t) \, dt \right], \qquad 0 \le x \le b,$$

where $u : \mathbb{R} \to \mathbb{R}$ is bounded and measurable. Use the strong Markov property of X to show that

$$k(x) = h(x) + \psi_1(x)k(q) + \psi_2(x)k(Q)$$

where $h(\cdot)$, $\psi_1(\cdot)$, and $\psi_2(\cdot)$ are as computed earlier in Section 3.3 and Section 3.6. Then show that $k(q)$ and $k(Q)$ are uniquely determined by the vector-matrix relation

$$\begin{bmatrix} k(q) \\ k(Q) \end{bmatrix} = \begin{bmatrix} h(q) \\ h(Q) \end{bmatrix} + \begin{bmatrix} \psi_1(q) & \psi_2(q) \\ \psi_1(Q) & \psi_2(Q) \end{bmatrix} \begin{bmatrix} k(q) \\ k(Q) \end{bmatrix}.$$

4

Stochastic Calculus

It is the purpose of this chapter to state, in a form suitable for later applications, several variations of the Itô differentiation formula for Brownian motion. Because we seek to record only such information as is required for intelligent application of the Itô calculus, only a few propositions will be proved; see Chapter 3 of Karatzas and Shreve (1998) for a complete mathematical development. Nonetheless, it is a substantial task to just *state* the results of interest in precise mathematical terms. In particular, we must define what is meant by integration with respect to Brownian motion.

4.1 Introduction

Departing from previous usage, let us denote by W a standard Brownian motion (or Wiener process) on some filtered probability space $(\Omega, \mathcal{F}, \mathbb{F}, P)$. Readers should review the meaning given to this phrase in Section 1.1 (specifically, in the last two paragraphs), recalling in particular that

$$(4.1) \qquad W(t + u) - W(t) \text{ is independent of } \mathcal{F}_t \text{ for all } t, u \geq 0.$$

Our objective in Sections 4.2 to 4.4 is to define a continuous stochastic process

$$(4.2) \qquad I_t(X) := \int_0^t X \, dW, \qquad t \geq 0,$$

for a certain class of processes X. To be specific, let H denote the set of all integrands X such that

$$(4.3) \qquad X \text{ is an adapted process on } (\Omega, \mathbb{F}, P)$$

and

$$(4.4) \qquad P\left\{ \int_0^t X^2(s) \, ds < \infty \right\} = 1 \qquad \text{for all } t \geq 0.$$

We seek to give (4.2) a precise meaning for all $X \in H$. The random variables $I_t(X)$ will be called *stochastic integrals* (of X with respect to W), and the entire process $\{I_t(X), \ t \geq 0\}$ will be denoted by the indefinite integral $\int X \, dW$.

Remember that we only use the term *stochastic process* when referring to functions $X(\omega, t)$ that are jointly measurable in ω and t (see Section A.2) and thus joint measurability is implicit in (4.3). Conditions (4.1) and (4.3) together imply that $\{X(s), \ 0 \leq s \leq t\}$ and $\{W(t + u) - W(t), \ u \geq 0\}$ are independent for each t. This restriction is essential for the theory developed here, and the interested reader may consult page 31 of McKean (1969) to see why (4.4) is indispensable as well.

Suppose that X is a VF process, meaning that almost every sample path is a VF function in the sense of Section B.2. Then Theorem B.4 shows that $I_t(X)$ can be defined for each fixed ω in the Riemann–Stieltjes sense. Unfortunately, our primary interest is in the case where X has unbounded variation like W; hence the stochastic integral cannot be defined in any conventional sense. What will be shown in Sections 4.2 to 4.4 is that $I_t(X)$ can be defined for integrands $X \in H$ by a limiting procedure that is probabilistically natural and intuitive.

To focus attention on the basic concepts, we begin by defining $I_t(X)$ for individual time points t and integrands X that satisfy a stronger condition than (4.4). This definition is actually quite simple, and most of the steps will be proved. Section 4.3 is devoted to analysis of a revealing example. The general definition of the stochastic interval is laid out in Section 4.4, and stochastic differential equations are introduced in Section 4.5. The basic Itô formula and various generalizations are presented in Sections 4.6 to 4.10, and Section 4.11 gives some first applications.

4.2 First definition of the stochastic integral

Hereafter let H^2 be the set of all adapted processes X on $(\Omega, \mathcal{F}, \mathbb{F}, P)$ satisfying

$$(4.5) \qquad E\left[\int_0^t X^2(s) \, ds\right] < \infty \qquad \text{for all } t \geq 0.$$

Condition (4.5) is stronger than (4.4) and so H^2 is a proper subset of H. A process X will be called *simple* if there exist times $\{t_k\}$ such that

$$(4.6) \qquad 0 = t_0 < t_1 < \cdots < t_k \longrightarrow \infty$$

and

(4.7) $X(t, \omega) = X(t_{k-1}, \omega)$ for all $t \in [t_{k-1}, t_k)$ and $k = 1, 2, \ldots$.

Note that the times $\{t_k\}$ do not depend on ω. Let S be the set of all simple adapted processes, and let S^2 be the set of all simple $X \in H^2$.

Let L^2 denote as usual the set of all random variables ξ on $(\Omega, \mathcal{F}, P)$ such that

(4.8) $\|\xi\| := \left[E(\xi^2) \right]^{1/2} < \infty.$

When we say that $\xi_n \to \xi$ in L^2, this means that $\xi_1, \xi_2, \ldots, \xi$ are all elements of L^2 and $\|\xi_n - \xi\| \to 0$. A sequence $\{\xi_n\}$ in L^2 is said to be *fundamental* (or a Cauchy sequence) if $\|\xi_m - \xi_n\|$ can be made arbitrarily small by taking m and n sufficiently large. The following is a well-known result from analysis.

Proposition 4.1 L^2 *is complete. That is, every fundamental sequence has a limit in* L^2.

Fixing $t > 0$ until further notice, we now define $I_t(X)$ for $X \in H^2$. To begin, let

(4.9) $\|X\| := E\left[\int_0^t X^2(s)\, ds \right]^{1/2}$ for $X \in H^2$.

Although the same symbol $\| \cdot \|$ is used to denote a norm on L^2 and a norm on H^2, attentive readers will find that this causes no serious confusion. Convergence of sequences in H^2 is defined just as for sequences in L^2. The following proposition is important but rather technical, so its proof is omitted.

Proposition 4.2 S^2 *is dense in* H^2. *That is, for each* $X \in H^2$ *there exist simple processes* $\{X_n\}$ *such that*

(4.10) $X_n \longrightarrow X \in H^2$ *as* $n \to \infty$.

To simplify notation, set $I(X) := I_t(X)$ until t is freed. If X is simple, then one can define $I(X)$ in the Riemann–Stieltjes sense (see Section B.3) for almost every ω. To be specific, let us suppose (without loss of generality) that (4.6) and (4.7) hold with $t = t_n$. Then the Riemann–Stieltjes theory defines

(4.11) $I(X) = \sum_{k=0}^{n-1} X(t_k)\, [W(t_{k+1}) - W(t_k)].$

Remember that $I(X)$ is a random variable, although we suppress its dependence on ω in the usual way.

Proposition 4.3 *If $X \in S^2$, then $E[I(X)] = 0$ and $\|I(X)\| = \|X\|$.*

Remark The second part of the conclusion says that $I(X) \in L^2$ and that the L^2 norm of $I(X)$ equals the H^2 norm of X.

Proof Again suppose (4.6) and (4.7) hold with $t = t_n$ and write $\mathcal{F}_k$ in place of $\mathcal{F}(t_k)$ to simplify notation. For the first part, we use (4.11), (4.1), and the adaptedness of X to write

$$E[I(X)] = \sum_{k=0}^{n-1} E\{X(t_k)[W(t_{k+1}) - W(t_k)]\}$$

$$= \sum_{k=0}^{n-1} E\{E\{X(t_k)[W(t_{k+1}) - W(t_k)]|\mathcal{F}_k\}\}$$

$$= \sum_{k=0}^{n-1} E\{X(t_k)E[W(t_{k+1}) - W(t_k)|\mathcal{F}_k]\} = 0.$$

For the second part, note that

$$I^2(X) = \sum_{k=0}^{n-1} X^2(t_k)[W(t_{k+1}) - W(t_k)]^2$$

$$+ 2\sum_{j=0}^{n-2}\sum_{k=j+1}^{n-1} X(t_j)X(t_k)[W(t_{j+1}) - W(t_j)][W(t_{k+1}) - W(t_k)].$$

Now take the expectation of each side, first conditioning on $\mathcal{F}_k$ in the kth term of the first sum and in the (j,k)th term of the second sum. The first three factors inside the double sum are all measurable with respect to $\mathcal{F}_k$, whereas the conditional expectation of the last factor given $\mathcal{F}_k$ is zero as above. Thus the double sum has zero expectation and we come down to

$$E[I^2(X)] = E\left[\sum_{k=0}^{n-1} X^2(t_k)E\{[W(t_{k+1}) - W(t_k)]^2|\mathcal{F}_k\}\right]$$

$$= E\left[\sum_{k=0}^{n-1} X^2(t_k)(t_{k+1} - t_k)\right] = E\left[\int_0^t X^2(s)\,ds\right] = \|X\|^2.$$

Because $E[I^2(X)] = \|I(X)\|^2$, this completes the proof. $\qquad\square$

Proposition 4.4   *Suppose that $X \in H^2$. There exists a random variable $I(X) \in L^2$, unique up to an equivalence, such that $I(X_n) \to I(X)$ in L^2 for each simple sequence $\{X_n\}$ satisfying (4.10). Furthermore, $E[I(X)] = 0$ and $\|I(X)\| = \|X\|$.*

Remark   The phrase "unique up to an equivalence" means that any two random variables fitting this description are equal almost surely. Combining Proposition 4.2 and Proposition 4.4, the stochastic integral $I(X)$ is defined up to an equivalence for each $X \in H^2$.

Proof   Let $\{X_n\}$ be a sequence of simple processes for which (4.10) holds. Then $\{X_n\}$ is a fundamental sequence in H^2. Proposition 4.3 gives

$$\|I(X_m) - I(X_n)\| = \|I(X_m - X_n)\| = \|X_m - X_n\|$$

and hence $\{I(X_n)\}$ is a fundamental sequence in L^2. Thus by Proposition 4.1 there exists some random variable $I(X) \in L^2$ such that $I(X_n) \to I(X)$. If $\{X'_n\}$ is any other sequence in S^2 for which (4.10) holds, then $\|X_n - X'_n\| \le \|X_n - X\| + \|X - X'_n\| \to 0$, and another application of Proposition 4.3 gives

$$\|I(X'_n) - I(X_n)\| = \|I(X'_n - X_n)\| = \|X'_n - X_n\| \longrightarrow 0.$$

Thus $\|I(X'_n) - I(X)\| \le \|I(X'_n) - I(X_n)\| + \|I(X_n) - I(X)\| \to 0$, meaning that $I(X'_n) \to I(X)$ as well. This establishes the uniqueness of the stochastic integral $I(X)$. If $\xi_n \to \xi$ in L^2, it is well known (and an easy consequence of the dominated convergence theorem) that $E(\xi_n) \to E(\xi)$ and $\|\xi_n\| \to \|\xi\|$. The last statement of Proposition 4.4 follows from this and from Proposition 4.3. Thus the proof is complete.   □

4.3 An illuminating example

We now consider the integrand $X = W$, seeking to compute $I_t(W) := \int_0^t W \, dW$ explicitly. Using Fubini's theorem (see Section A.5), one notes first that

$$(4.12) \qquad E\left[\int_0^t W^2(s) \, ds\right] = \int_0^t E\left[W^2(s)\right] ds = \int_0^t s \, ds = \frac{t^2}{2} < \infty$$

and thus $W \in H^2$. Now fix $t > 0$ and consider the simple functions $\{X_n\}$ defined by

$$(4.13) \qquad X_n(s) := W\left(\frac{kt}{2^n}\right) \qquad \text{for } s \in \left[\frac{kt}{2^n}, \frac{(k+1)t}{2^n}\right)$$

and $k = 0, 1, \ldots$. Check that $X_1, X_2, \ldots$ are adapted and are furthermore elements of H^2. Next, defining the H^2 norm as in Section 4.2,

$$
\begin{aligned}
\|W - X_n\| &= E\left\{ \int_0^t [W(s) - X_n(s)]^2 \, ds \right\} \\
&= \int_0^t E\left\{ [W(s) - X_n(s)]^2 \right\} ds \\
(4.14) \qquad &= \sum_{k=0}^{2^n-1} \int_0^{t/2^n} E\left\{ \left[W\left(\frac{kt}{2^n} + s \right) - W\left(\frac{kt}{2^n} \right) \right]^2 \right\} ds \\
&= \sum_{k=0}^{2^n-1} \int_0^{t/2^n} s \, ds = 2^n \frac{1}{2} \left(\frac{t}{2^n} \right)^2 = \frac{t^2}{2^{n+1}}.
\end{aligned}
$$

Thus $\|W - X_n\| \to 0$, implying by Proposition 4.4 that $I_t(X_n) \to I_t(W)$. Fixing n for the moment, write t_k in place of $kt/2^n$. Specializing (4.11) to the simple processes X_n gives

$$
\begin{aligned}
I_t(X_n) &= \sum_{k=0}^{2^n-1} W(t_k) [W(t_{k+1}) - W(t_k)] \\
&= \frac{1}{2} \sum_{k=0}^{2^n-1} \left\{ \left[W^2(t_{k+1}) - W^2(t_k) \right] - \left[W^2(t_{k+1}) + W^2(t_k) \right] \right. \\
(4.15) \qquad &\qquad\qquad \left. + 2W(t_k) W(t_{k+1}) \right\} \\
&= \frac{1}{2} \sum_{k=0}^{2^n-1} \left[W^2(t_{k+1}) - W^2(t_k) \right] - \frac{1}{2} \sum_{k=0}^{2^n-1} [W(t_{k+1}) - W(t_k)]^2 \\
&= \frac{1}{2} W^2(t) - \frac{1}{2} \sum_{k=0}^{2^n-1} [W(t_{k+1}) - W(t_k)]^2 .
\end{aligned}
$$

In Section 1.2 (dealing with quadratic variation) it was shown that the summation in the last line (a random variable) converges in the L^2 sense to t. Combining all this gives

$$
(4.16) \qquad I_t(X_n) \longrightarrow \frac{1}{2} W^2(t) - \frac{1}{2} t \qquad \text{in the } L^2 \text{ sense}
$$

and consequently

$$
(4.17) \qquad \int_0^t W \, dW = \frac{1}{2} W^2(t) - \frac{1}{2} t.
$$

If W were a continuous VF process with $W(0) = 0$, formula (B.4) would give us $W^2(t) = 2 \int_0^t W \, dW$, with the integral defined in the Riemann–Stieltjes sense. Thus the surprising thing about (4.17) is the term $-t/2$ on the right. This peculiarity traces to the infinite variation of Brownian paths, and more particularly to their positive quadratic variation. Consider now simple processes $\{X'_n\}$ defined by

$$(4.18) \qquad X'_n(s) = W\left(\frac{(k+1)t}{2^n}\right) \qquad \text{for } s \in \left[\frac{kt}{2^n}, \frac{(k+1)t}{2^n}\right)$$

and $k = 0, 1, \ldots$. In contrast with (4.13), this scheme approximates W over each interval $[kt/2^n, (k+1)t/2^n)$ by its value at the *right* endpoint. If X'_n is substituted for X_n in (4.15), one ultimately finds that $I_t(X'_n) \to \frac{1}{2} W^2(t) + t/2$ as $n \to \infty$. If W were Riemann–Stieltjes integrable with respect to itself, the substitution of $\{X'_n\}$ for $\{X_n\}$ would make no difference as $n \to \infty$, but we find that this substitution does make a difference when W is a Brownian motion. The key point here is that the simple processes $\{X'_n\}$ are *not* adapted and hence are not elements of S^2; they cannot be used in approximating $I_t(W)$. The approximating simple functions $\{X_n\}$ *are* elements of S^2 and hence (4.17) gives the correct value of $I_t(W)$ in Itô's stochastic calculus.

4.4 Final definition of the integral

In Section 4.2 we defined the stochastic integral $I_t(X)$ for $X \in H^2$ and a fixed time $t \geq 0$, suppressing the subscript t to simplify notation. The subscript is restored for the following proposition, which can be verbally paraphrased as follows: Proposition 4.4 associates with each time $t \geq 0$ an equivalence class of L^2 random variables $I_t(X)$, and one can select a member of that class for each time t in such a way that $\{I_t(X), \ t \geq 0\}$ is a continuous process.

Proposition 4.5 *Fix $X \in H^2$. There exists a continuous process $Z = \{Z_t, \ t \geq 0\}$, unique up to an equivalence, such that $P\{Z_t = I_t(X)\} = 1$ for each fixed $t \geq 0$, where $I_t(X)$ is the L^2 limit of simple approximations identified in Proposition 4.4.*

Remark In this context the phrase "unique up to an equivalence" means if Z and Z^* are any two such processes, then $P(Z_t = Z_t^*$ for all $t \geq 0) = 1$.

Finally, we extend the definition of the stochastic integral from integrands $X \in H^2$ to integrands $X \in H$ by a process called *localization*, as follows. Fixing $X \in H$, let

$$T_n := \inf \left\{ t \geq 0 : \int_0^t X^2(s)\, ds \geq n \right\}, \qquad n \geq 1,$$

with $T_n := \infty$ if the indicated t-set is empty. Then each T_n is a stopping time, and one has

$$P(0 < T_1 \leq T_2 \leq \cdots \to \infty) = 1$$

from the definition of H. Next, defining

$$X_n(t) := \begin{cases} X(t) & \text{if } 0 \leq t \leq T_n \\ 0 & \text{if } t > T_n, \end{cases}$$

observe that $X_n \in H^2$ and hence one can define a continuous stochastic process $\{I_t(X_n),\ t \geq 0\}$ via Proposition 4.5 for each $n = 1, 2, \ldots$. But $X_n, X_{n+1}, \ldots$ all coincide over the interval $[0, T_n]$, and because $T_n \to \infty$ almost surely one has the following: there exists a continuous process $Z = \{Z_t,\ t \geq 0\}$ such that

(4.19) $\qquad P\{I_t(X_n) \to Z(t) \text{ for each } t \geq 0 \text{ as } n \uparrow \infty\} = 1.$

Definition 4.6 When we refer to $\int X\, dW$ hereafter, this is understood to mean the continuous process Z in (4.19).

From the definitions above it follows that $\int (aX_1 + bX_2)\, dW = a \int X_1\, dW + b \int X_2\, dW$; this linearity will be used often without further comment. The following proposition, which generalizes the last part of Proposition 4.4, gives the only other property of the integral we shall need. Its proof is left as an exercise (see Problem 4.1).

Proposition 4.7 *Suppose $X \in H$, $Z = \int X\, dW$, and T is a stopping time. If*

(4.20) $$E\left[\int_0^T X^2(t)\, dt \right] < \infty$$

then $E[Z(T)] = 0$ and $E[Z^2(T)] = E\left[\int_0^T X^2(t)\, dt \right].$

Corollary 4.8 *If $\{X(t),\ 0 \leq t \leq T\}$ is bounded and $E(T) < \infty$, then $E[Z(T)] = 0$.*

Both Proposition 4.7 and Corollary 4.8 will be referred to later as *zero expectation properties* of the stochastic integral. A more fundamental property is that $\int X\, dW$ is a martingale for $X \in H^2$ and is what is called a *local martingale* for all $X \in H$.

4.5 Stochastic differential equations

The term "stochastic differential equation" (SDE) is used here to mean a relationship of the form

$$(4.21) \qquad dX(t) = \mu(X(t))\, dt + \sigma(X(t))\, dW(t),$$

which is intended to define a process X in terms of a standard Brownian motion W. The data of the process X are the coefficient functions $\mu(\cdot)$ and $\sigma(\cdot)$, plus an initial state $X(0)$ that is taken to be deterministic throughout this section. The precise meaning of (4.21) is that

$$(4.22) \quad X(t) = X(0) + \int_0^t \mu(X(s))\, ds + \int_0^t \sigma(X(s))\, dW(s), \qquad t \ge 0,$$

with the stochastic integral on the right side of (4.22) defined as in the preceding section. Intuitively, this means that over any small time interval $[t, t + \Delta t]$ one has

$$(4.23) \qquad \Delta X(t) \simeq \mu(X(t))\, \Delta t + \sigma(X(t))\, \Delta W(t),$$

where $\Delta X(t) := X(t + h) - X(t)$ and similarly for $\Delta W(t)$. That is, the conditional distribution of $\Delta X(t)$ given the history of W (and hence also of X) up to time t is approximately Gaussian with mean $\mu(X(t))\Delta t$ and with variance $\sigma^2(X(t))\Delta t$. To put that another way, the process X behaves locally like a (μ, σ) Brownian motion, but with the drift and variance parameters depending on the current state of X through the coefficient functions $\mu(\cdot)$ and $\sigma(\cdot)$.

In Chapters 5, 8, and 9, readers will see examples of processes which can be defined initially without reference to stochastic integrals or SDE theory, and are then shown to satisfy a relationship of the form (4.22). In general, though, it is natural to ask whether there exists a process X that satisfies (4.22), and if so, whether it is unique. The answer is affirmative if the coefficient functions are sufficiently regular. That general theory is not needed for the applications considered in this book, so we shall say no more on the subject, referring interested readers to Chapter 5 of Øksendal (2007) or Chapter 5 of Karatzas and Shreve (1998).

Processes of the form (4.22) are called (one-dimensional) *Itô diffusion processes*, or *Itô diffusions* (the historical rationale for the word "diffusion" was explained in Section 3.2), but the modifier "Itô" will be dropped in the remainder of this book. That is, X will be called simply a "diffusion process" or a "diffusion." For a glimpse of the associated analytical theory see Problem 4.18.

4.6 Simplest version of Itô's formula

We continue with the setup where W is a standard Brownian motion on the filtered probability space $(\Omega, \mathcal{F}, \mathbb{F}, P)$. The term *Itô process* will be used here to mean a process Z that is representable in the form

$$(4.24) \qquad Z(t) = Z(0) + \int_0^t X \, dW + A(t), \qquad t \geq 0,$$

where $Z(0) \in \mathcal{F}_0$, $X \in H$, and A is a continuous, adapted VF process with $A(0) = 0$. Thus Z is itself continuous and adapted. (Actually, the term *Itô process* will be used later in a slightly broader sense, but this will cause no confusion.) Our definition is a bit more generous than usual; most writers impose the further requirement that A be absolutely continuous, meaning that

$$(4.25) \qquad A(\omega, t) = \int_0^t Y(\omega, s) \, ds, \qquad t \geq 0,$$

where Y is jointly measurable in t and ω, is adapted, and satisfies

$$(4.26) \qquad \int_0^t |Y(s)| \, ds < \infty \text{ almost surely}, \qquad t \geq 0.$$

All important results with the usual definition carry over to our more general setting, and the resulting gain is important for our purposes. Specifically, reflected Brownian motion is an Itô process according to our definition but not according to the standard one (see Section 1.10).

Hereafter, the second term on the right side of (4.24) will be called the *Brownian component* of Z, and A will be called the *VF component*. When we say that a process Z has an *Itô differential* $X \, dW + dA$, or simply write $dZ = X \, dW + dA$, this is understood to be shorthand for the precise statement (4.24), and it is similarly understood that X and A meet the restrictions stated immediately after (4.24). Also, when we say that Z is an Itô process with differential $dZ = X \, dW + Y \, dt$, this is understood as shorthand for (4.24) and (4.25) together, with X and Y meeting all the necessary restrictions. Incidentally, when $dZ = X \, dW + Y \, dt$, the VF process $A = \int Y(s) \, ds$ is usually called the *drift component* of Z, but that terminology will not be used here.

We now give an exact and explicit statement of the Itô differentiation rule in its simplest form, followed by several equivalent statements of the rule that are less explicit but more compact. A sketch of the proof will then be given. For a process X and function $\phi : \mathbb{R} \to \mathbb{R}$, we shall hereafter denote by $\phi(X)$ the entire process $\{\phi(X_t), \ t \geq 0\}$.

Proposition 4.9   *Suppose that $f : \mathbb{R} \to \mathbb{R}$ is C^2 and Z is an Itô process with $dZ = X\,dW + dA$. Then*

$$(4.27) \quad f(Z_t) = f(Z_0) + \int_0^t [f'(Z)X]\,dW$$

$$+ \int_0^t f'(Z)\,dA + \frac{1}{2} \int_0^t \left[f''(Z)X^2 \right] ds, \qquad t \geq 0,$$

where the first integral on the right is defined as in Section 4.4, the second is defined path by path in the Riemann–Stieltjes sense (see Section B.3), and the third is defined path by path in the Riemann sense.

First Remark   It is customary to express (4.27) more compactly as

$$(4.28) \qquad\qquad df(Z) = f'(Z)\,dZ + \tfrac{1}{2}f''(Z)X^2\,dt$$

with the following conventions understood. First, of course, is the fact that any equation involving Itô differentials is shorthand for a precise statement in terms of stochastic integrals. Second, dZ is shorthand for $X\,dW + dA$ and we separate the dW and dA terms that result from this substitution. Thus (4.28) can be written more explicitly as

$$df(Z) = f'(Z)X\,dW + f'(Z)\,dA + \tfrac{1}{2}f''(Z)X^2\,dt.$$

Second Remark   An even more highly symbolic expression of (4.27), and one that has real advantages in its multi-dimensional form, is

$$(4.29) \qquad\qquad df(Z) = f'(Z)\,dZ + \tfrac{1}{2}f''(Z)(dZ)^2.$$

Here it is understood that one computes $(dZ)^2$ as

$$(4.30) \qquad (dZ)^2 = (X\,dW + dA)^2 = X^2(dW)^2 + 2X\,dW\,dA + (dA)^2$$

and then computes the various products according to Table 4.1. That is, only the first term on the right side of (4.30) is nonzero, and its value is $X^2\,dt$, so (4.29) reduces to (4.28).

Table 4.1 *Multiplication table for products of $(dZ)^2$.*

	dW	dA
dW	dt	0
dA	0	0

Third Remark   We saw in Section B.4 that the Riemann–Stieltjes calculus yields $df(Z) = f'(Z) dZ$ if Z is a continuous VF process. So the novel feature of (4.28) or (4.29) is the second term on the right side. It will be seen shortly that this traces to the positive quadratic variation of Brownian paths. Also, the following example connects the mysterious second term with our earlier surprising discovery that $2 \int_0^t W \, dW = W^2(t) - t$ in the Itô calculus (see Section 4.3). If $Z = W$ and $f(x) = x^2$, then (4.28) gives $dW^2 = 2W \, dW + dt$. In precise integral form, this says that $W^2(t) = 2 \int_0^t W \, dW + t$.

Sketch of Proof   The traditional method of proof uses Taylor's theorem. For the special case $Z = W$, the argument goes as follows. Fix $t > 0$ and let $0 = t_0 < \cdots < t_n = t$ be a partition of $[0, t]$. Then

$$(4.31) \qquad f(W_t) - f(0) = \sum_{k=0}^{n-1} [f(W(t_{k+1})) - f(W(t_k))].$$

According to Taylor's theorem (with the exact form of the remainder), each term on the right can be written as

$$(4.32) \quad f(W(t_{k+1})) - f(W(t_k)) = f'(W(t_k))[W(t_{k+1}) - W(t_k)]$$
$$+ \frac{1}{2} f''(\xi_k)[W(t_{k+1}) - W(t_k)]^2$$

where ξ_k lies in the interval between $W(t_k)$ and $W(t_{k+1})$. Because W is continuous we can then write $\xi_k(\omega) = W(\tau_k(\omega))$, where $t_k \leq \tau_k(\omega) \leq t_{k+1}$. Thus (4.31) becomes

$$(4.33) \quad f(W_t) - f(0) = \sum_{k=0}^{n-1} f'(W(t_k))[W(t_{k+1}) - W(t_k)]$$
$$+ \frac{1}{2} \sum_{k=0}^{n-1} f''(W(\tau_k))[W(t_{k+1}) - W(t_k)]^2.$$

Note that the first term on the right is the stochastic integral of a simple adapted process that closely approximates $f'(W)$ if the partition is fine. Also the quadratic variation theorem of Section 1.2 suggests that the second sum on the right side of (4.33) will be closely approximated by

$$\sum_{k=0}^{n-1} f''(W(\tau_k))(t_{k+1} - t_k)$$

if the partition is fine. Thus the following statement is unsurprising: there exists a sequence of successively finer partitions for which

$$(4.34) \qquad \sum_{k=0}^{n-1} f'(W(t_k)) [W(t_{k+1}) - W(t_k)] \longrightarrow \int_0^t f'(W) \, dW$$

and

$$(4.35) \qquad \sum_{k=0}^{n-1} f''(W(\tau_k)) [W(t_{k+1}) - W(t_k)]^2 \longrightarrow \int_0^t f''(W) \, ds,$$

both statements holding in the almost sure sense. Substituting (4.34) and (4.35) in (4.33) gives

$$f(W_t) - f(0) = \int_0^t f'(W) \, dW + \frac{1}{2} \int_0^t f''(W) \, ds, \qquad t \ge 0,$$

which is the specialization of (4.27) to the case under discussion.

4.7 The multi-dimensional Itô formula

Let us now generalize our setup, assuming that $W_1, \ldots, W_n$ are n *independent* Wiener processes on the filtered probability space $(\Omega, \mathcal{F}, \mathbb{F}, P)$. We consider here a vector process $Z = (Z_1, \ldots, Z_m)$ whose components can be represented in the form

$$(4.36) \qquad Z_i(t) = Z_i(0) + \sum_{j=1}^{n} \int_0^t X_{ij} \, dW_j + A_i(t), \qquad t \ge 0,$$

where $Z_i(0)$ is measurable with respect to $\mathcal{F}_0$, $X_{ij} \in H$, and A_i is a continuous, adapted VF process. This is the general form of a *multi-dimensional Itô process*. Note that each of the stochastic integrals on the right side of (4.36) is well defined by the development in Section 4.4 so there is nothing new as yet. In the obvious way, we write

$$(4.37) \qquad dZ_i = \sum_{j=1}^{n} X_{ij} \, dW_j + dA_i \qquad (i = 1, \ldots, m)$$

as shorthand for (4.36). Assuming that the precise meaning of differential statements like (4.37) is now clear, we shall state the multi-dimensional Itô formula only in the compact symbolic form analogous to (4.29).

Proposition 4.10 *Suppose that $f : \mathbb{R}^m \to \mathbb{R}$ is twice continuously differentiable, meaning that all the first-order partials f_i and second-order* partials f_{ij} exist and are continuous. If Z satisfies (4.37), then $f(Z)$ is an Itô process with differential

$$(4.38) \qquad df(Z) = \sum_{i=1}^{m} f_i(Z)\, dZ_i + \frac{1}{2} \sum_{i=1}^{m} \sum_{k=1}^{m} f_{ik}(Z)\, dZ_i\, dZ_k$$

where the products $dZ_i\, dZ_k$ are computed using (4.37) and the multiplication rules $dW_j\, dW_k = \delta_{jk}\, dt$, $dW_j\, dA_i = 0$, $dA_i\, dA_k = 0$. (Here $\delta_{jk} = 1$ if $j = k$ and $= 0$ otherwise.)

Formula (4.38) is analogous to (4.29) in its level of symbolism. To assure that the multiplication rule is clearly understood, readers should verify that (4.38) becomes

$$(4.39) \quad df(Z) = \sum_{i=1}^{m} \sum_{j=1}^{n} \left[f_i(Z)X_{ij} \right] dW_j + \sum_{i=1}^{m} f_i(Z)\, dA_i$$

$$+ \frac{1}{2} \sum_{i=1}^{m} \sum_{k=1}^{m} \sum_{j=1}^{n} \left[f_{ik}(Z)X_{ij}X_{kj} \right] dt$$

upon substitution of (4.37) and simplification. This version of the multi-dimensional formula is analogous to (4.28). Assuming that the exact meaning of such differential statements is clear from Section 4.6, we shall not write out (4.39) any more explicitly. In the future we shall consistently use the symbolism of (4.38) both because of its compactness and because this version of the formula is so much easier to remember than (4.39), at least for those familiar with the multi-dimensional Taylor formula.

Proposition 4.10 says that a smooth function f of an Itô process Z is itself an Itô process, with differential given by (4.38). What if we form a new process Z^* via $Z_t^* := \phi(Z_t, t)$? If ϕ is twice continuously differentiable as a function of $m+1$ variables, then this situation is covered by Proposition 4.10. Furthermore, all the conclusions of Proposition 4.10 remain valid with slightly weaker assumptions on ϕ; the *second-order* partials involving t need not be continuous or even exist.

4.8 Tanaka's formula and local time

If Z is a one-dimensional Itô process and f is twice continuously differentiable, we have seen in Section 4.6 that $f(Z)$ is also an Itô process, and its differential is given by (4.28). What if f is not so smooth? In this section we address that question for the special case where Z is a (μ, σ) Brownian motion on $(\Omega, \mathcal{F}, \mathbb{F}, P)$. To introduce the basic ideas in a simple setting, consider first the case $f(x) = |x|$. More precisely, in an effort to approximate the absolute value by a smoother function, let $t > 0$ be arbitrary and define $f : \mathbb{R} \to \mathbb{R}$ via

(4.40) $f(0) = f'(0) = 0$

and

(4.41) $f''(x) = \begin{cases} 1/\epsilon & \text{if } |x| \le \epsilon \\ 0 & \text{otherwise.} \end{cases}$

It follows that

(4.42) $f'(x) = \begin{cases} x/\epsilon & \text{if } |x| \le \epsilon \\ \operatorname{sgn}(x) & \text{otherwise} \end{cases}$

and

(4.43) $f(x) = \begin{cases} x^2/2\epsilon & \text{if } |x| \le \epsilon \\ |x| - \frac{1}{2}\epsilon & \text{otherwise.} \end{cases}$

Figure 4.1 shows f and its first two derivatives. If f had a *continuous* second derivative, then we could apply the basic Itô formula (4.27) to obtain

(4.44) $f(Z_t) = f(Z_0) + \int_0^t f'(Z)\,dZ + \frac{1}{2}\sigma^2 \int_0^t f''(Z)\,ds.$

Furthermore, (4.44) remains valid for the function f defined by (4.43), as can be proved with an approximation argument. Substituting (4.41) into the last term of (4.44) and denoting by $l(t, x)$ the local time of Z at level x (see Section 1.2), we have

(4.45) $\frac{1}{2} \int_0^t f''(Z)\,ds = \frac{1}{2\epsilon} \int_0^t 1_{\{|Z(s)| \le \epsilon\}}\,ds \longrightarrow l(t, 0) \qquad \text{as } \epsilon \downarrow 0.$

Furthermore, using the explicit formula (4.42), it is not difficult to show that the second term on the right side of (4.44) approaches $\int \operatorname{sgn}(Z)\,dZ$ as $\epsilon \downarrow 0$. Of course, $f(Z_t) \to |Z_t|$ as $\epsilon \downarrow 0$ by (4.43). Combining this with (4.44) and (4.45), we have

(4.46) $|Z_t| = \int_0^t \operatorname{sgn}(Z)\,dZ + \sigma^2 l(t, 0), \qquad t \ge 0.$

In the particular case where Z is a *standard* Brownian motion, (4.46) is called *Tanaka's formula*. To generalize further, let us introduce the following. (We write RCLL to indicate a right-continuous function for which left limits exist.)

Assumption 4.11 Let $f : \mathbb{R} \to \mathbb{R}$ be absolutely continuous with RCLL density f'. It is assumed that f' has finite variation on every finite interval. Let ρ be the signed measure on $(\mathbb{R}, \mathcal{B})$ defined by $\rho(a, b] = f'(b) - f'(a)$ for $-\infty < a < b < \infty$.

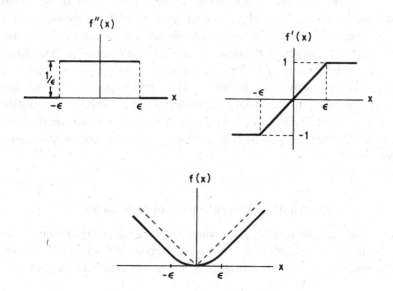

Figure 4.1 Approximating the absolute value function.

One may paraphrase Assumption 4.11 by saying that the second derivative of f exists as a measure. A function satisfies this description if and only if it can be written as the difference of two convex functions. The following is proved in Section 9.2 of Chung and Williams (1990) for the case of standard Brownian motion, and the extension to general drift and variance parameters is trivial.

Proposition 4.12 *If Z is a (μ, σ) Brownian motion and f satisfies Assumption 4.11 then*

$$(4.47) \qquad f(Z_t) = f(Z_0) + \int_0^t f'(Z)\, dZ + \frac{1}{2}\sigma^2 \int_{\mathbb{R}} l(t, x)\, \rho(dx).$$

If f is C^2 then $\rho(dx) = f''(x)\,dx$, and equation (1.7) specializes to give

$$(4.48) \qquad \int_{\mathbb{R}} l(t, x)\rho(dx) = \int_{\mathbb{R}} l(t, x)f''(x)\,dx = \int_0^t f''(Z)\,ds$$

and thus (4.47) reduces to the basic Itô formula (4.27) as it should. If f is C^1 and piecewise C^2 (see Guide to Notation and Terminology for the precise meaning of that term) then both equalities in (4.48) remain valid, with $f''(x)$ defined arbitrarily at points x where f'' is discontinuous. Thus the basic Itô formula extends to that situation as well. On the other hand, Proposition 4.12 shows that fundamentally new effects enter if f' has discontinuities. In particular, the right side of (4.47) has a term involving $l(t, x)$ for each point x where f' jumps. In the case where $f(x) = |x|$, $\rho\{0\} = 2$ and $\rho(dx) = 0$ away from the origin. Thus $\int l(t, x)\rho(dx) = 2l(t, 0)$, and (4.47) specializes to the Tanaka formula (4.46) as it should. Proposition 4.12 shows that $f(Z)$ is an Itô process in the sense that we use the term here (see Section 4.6) for any function f of the class described in Assumption 4.11.

4.9 Another generalization of Itô's formula

Having studied the process $f(Z)$ when f is less smooth than required by the basic Itô formula, let us now see what happens if Z is less smooth than assumed in Section 4.6. In this section, Z is assumed to have the form

$$(4.49) \qquad Z_t = \int_0^t X\,dW + V_t, \qquad t \ge 0,$$

where $X \in H$ and

$$(4.50) \qquad V \text{ is an adapted VF process.}$$

(Recall from Section A.2 that, throughout this book, the term "process" automatically implies right-continuous sample paths.) It follows that the left limit $V(t-)$ exists for all $t > 0$ almost surely and that V has just countably many points of discontinuity (or jumps) almost surely. Let us denote by $\Delta V(t) := V(t) - V(t-)$ the jump of V at time t, and define a new process A via

$$(4.51) \qquad A_t := V_t - \sum_{0 < s \le t} \Delta V_s, \qquad t \ge 0,$$

where the sum is over the countable set of $s \in (0, t]$ at which $|\Delta V_s| > 0$. Incidentally, because V has VF sample paths, we know that

$$\sum_{0<s\leq t} |\Delta V_s| < \infty$$

almost surely, so the sum in (4.51) makes sense. Obviously A is a continuous VF process; we call it the *continuous part* of V.

Proposition 4.13 *Suppose $f : \mathbb{R} \to \mathbb{R}$ is C^2 and Z satisfies (4.49) and (4.50). Then*

$$f(Z_t) = f(Z_0) + \int_0^t f'(Z)X\,dW + \int_0^t f'(Z)\,dA$$

$$+ \frac{1}{2}\int_0^t f''(Z)X^2\,ds + \sum_{0<s\leq t} \Delta f(Z)_s$$

where

$$\Delta f(Z)_s := f(Z_s) - f(Z_{s-}) \quad for\ s > 0.$$

If V (and hence Z) jumps only at isolated points $0 < T_1 < T_2 < \cdots \to \infty$, then Proposition 4.13 is just a trivial extension of the ordinary Itô formula; one can prove it by applying (4.27) on each of the intervals $[T_{n-1}, T_n)$ and adding. This observation can be combined with an approximation argument to prove Proposition 4.13 in general. The generalized Itô formula in Proposition 4.13 will play a major role in Chapter 7, where we study problems of optimal stochastic control.

4.10 Integration by parts (Special cases)

Let Y and Z be two Itô processes with differentials $dY = X\,dW + dA$ and $dZ = X^*\,dW + dA^*$, respectively. Note that Y and Z are built from a *common* standard Brownian motion W. Also, remember that our definition of Itô process requires that A and A^* be *continuous* VF processes. Let us apply the multi-dimensional Itô formula (4.38) to analyze the product $U_t := Y_t Z_t$. Defining $f(y, z) := yz$, we have

$$\frac{\partial}{\partial y}f = z, \quad \frac{\partial}{\partial z}f = y, \quad \frac{\partial^2}{\partial y^2}f = \frac{\partial^2}{\partial z^2}f = 0, \quad \frac{\partial^2}{\partial y\partial z}f = 1.$$

Of course, $U_t = f(y_t, Z_t)$, so (4.38) gives

(4.52) $$dU = Y\,dZ + Z\,dY + (dY)(dZ)$$

where

(4.53) $(dY)(dZ) = XX^* (dW)^2 = XX^* dt.$

Substituting (4.53) into (4.52) and writing out the precise integral form, we have

(4.54) $Y_t Z_t = Y_0 Z_0 + \displaystyle\int_0^t Y \, dZ + \int_0^t Z \, dY + \int_0^t XX^* \, ds.$

If *either* $X = 0$ or $X^* = 0$, meaning that either Y or Z is a VF process, then (4.54) reduces to the ordinary Riemann–Stieltjes integration by parts theorem (see Section B.3). The next proposition strengthens that statement slightly. The comments following Proposition 4.13 about method of proof apply equally here.

Proposition 4.14 *Suppose that* $Y = \int X \, dW + V$, *where* V *is an adapted VF process as in Section 4.9. If* Z *is a continuous VF process, then*

(4.55) $Y_t Z_t = Y_0 Z_0 + \displaystyle\int_0^t Y \, dZ + \int_0^t Z \, dY, \qquad t \geq 0.$

On the right side of (4.55), $\int Y \, dZ$ is interpreted as $\int (\int X \, dW) \, dZ + \int V \, dZ$. The integrand in the first term is continuous (an Itô process) and the integrand in the second term is a VF process and thus each integral can be defined path by path in the Riemann–Stieltjes sense of Proposition B.5. Similarly, $\int Z \, dY$ is interpreted as $\int ZX \, dW + \int Z \, dV$; the first term is a stochastic integral and the second is defined path by path in the Riemann–Stieltjes sense. By specializing Proposition 4.14 to the case $Z_t = \exp(-\lambda t)$, we get the following proposition, which will be used frequently in later discussion of expected discounted costs.

Proposition 4.15 *Let* Y *be as in Proposition 4.14. Then for any real constant* λ *and* $t \geq 0$ *we have*

$$e^{-\lambda t} Y_t = Y_0 + \int_0^t e^{-\lambda s} \, dY - \lambda \int_0^t e^{-\lambda s} Y \, ds.$$

4.11 Differential equations for Brownian motion

In Chapter 3 we used probabilistic methods to compute various interesting quantities associated with Brownian motion. After the fact, it was observed that the quantity in question, viewed as a function of starting state and perhaps time, satisfies a certain differential equation with certain boundary conditions. In this section it will be shown how Itô's formula can be used to derive such differential equations directly. Thus probabilistic questions can be recast in purely analytic terms and attacked with purely analytic methods. Some problems are most easily solved by such an approach or by a blend of probabilistic and analytic methods, as will be seen in the chapters that follow.

Given parameters μ and $\sigma > 0$, let $(\Omega, \mathcal{F}, \mathbb{F}, P_x)$ and X be as in Section 3.1. Thus X is a (μ, σ) Brownian motion with starting state x on the filtered probability space $(\Omega, \mathcal{F}, \mathbb{F}, P_x)$. Defining

$$W_t := \frac{1}{\sigma}(X_t - X_0 - \mu t), \qquad t \geq 0,$$

we observe that W is a standard Brownian motion on $(\Omega, \mathcal{F}, \mathbb{F}, P_x)$. Also, X is an Itô process with Brownian component σW and VF component μt. As in Chapter 3, we define the differential operator Γ via

$$\Gamma f := \tfrac{1}{2}\sigma^2 f'' + \mu f'.$$

Proposition 4.16 *If $f : \mathbb{R} \to \mathbb{R}$ is C^1 and piecewise C^2 then $f(X)$ is an Itô process with differential $df(X) = \sigma f'(X)\, dW + \Gamma f(X)\, dt$.*

Remark   One can define $f''(y)$ arbitrarily at those points y where $f''(\cdot)$ is discontinuous; see comments following Proposition 4.12.

Proof   First, as noted immediately after Proposition 4.12, the basic Itô formula (4.27) remains valid for a function f that satisfies the stated hypotheses. Thus we have

$$\begin{aligned}
df(X) &= f'(X)\, dX + \tfrac{1}{2}f''(X)\, (dX)^2 \\
&= f'(X)(\sigma\, dW + \mu\, dt) + \tfrac{1}{2}f''(X)\sigma^2\, dt \\
&= \sigma f'(X)\, dW + \Gamma f(X)\, dt. \qquad\qquad\qquad \square
\end{aligned}$$

Proposition 4.17   *Fixing $\lambda \geq 0$, let f be as in Proposition 4.16 and let $u := \lambda f - \Gamma f$. Let $a, b \in \mathbb{R}$ be such that $a < x < b$, and define the stopping time $T := T(a) \wedge T(b)$ as in Section 3.3. Then*

$$(4.56) \qquad f(x) = E_x\left[\int_0^T e^{-\lambda t}u(X_t)\, dt\right] + E_x\left[e^{-\lambda T}f(X_T)\right].$$

Remark   The fundamental identity (4.56) will be used in Problems 4.2 to 4.4 to verify certain calculations done earlier in Chapter 3.

Proof From Proposition 4.16 it is known that $df(X) = \sigma f'(X)\,dW + \Gamma f(X)\,dt$. Applying the specialized integration by parts formula of Proposition 4.15 with $f(X)$ in place of Y, one has

$$e^{-\lambda t} f(X_t) = f(X_0) + \int_0^t e^{-\lambda s}\,df(X) - \lambda \int_0^t e^{-\lambda s} f(X)\,ds$$

(4.57)
$$= f(X_0) + \sigma \int_0^t e^{-\lambda s} f'(X)\,dW$$

$$+ \int_0^t e^{-\lambda s}(\Gamma f - \lambda f)(X)\,ds.$$

Defining

(4.58)
$$M_t := \sigma \int_0^t e^{-\lambda s} f'(X)\,dW$$

we can rewrite (4.57) as

(4.59)
$$e^{-\lambda t} f(X_t) = f(X_0) + M_t - \int_0^t e^{-\lambda s} u(X_s)\,ds.$$

Because (4.59) is a sample path relationship (an almost sure equality between two continuous *processes*), it remains valid when T is substituted for t. Taking E_x of both sides then gives

(4.60)
$$E_x\left[e^{-\lambda T} f(X_T)\right] = f(x) + E_x(M_T) - E_x\left[\int_0^T e^{-\lambda t} u(X_t)\,dt\right].$$

The continuous function f' is bounded over $[a, b]$, so the integrand in (4.58) is bounded over the time interval $[0, T]$. It was shown in Section 3.3 that $E_x(T) < \infty$, so the zero expectation property of Corollary 4.8 gives $E_x(M_T) = 0$. When this is substituted in (4.60), the proof is complete. □

4.12 Problems and complements

Problem 4.1 Prove Proposition 4.7, using the following outline. First, defining $Y(s) := X(s)$ for $0 \le s \le T$ and $Y(s) := 0$ for $s > T$, observe that $Y \in H^2$ and thus $I_t(Y) := \int_0^t Y\,dW$ can be defined for each fixed t as in Section 4.2. Then Proposition 4.4 gives $E[I_t(Y)] = 0$ and

(4.61)
$$E[I_t^2(Y)] = E\left[\int_0^t Y^2(s)\,ds\right].$$

The right side of (4.61) increases monotonically to the finite limit (4.20) as $t \uparrow \infty$, from which it follows that the random variables $I_t(Y)$ converge in L^2

as $t \uparrow \infty$ to a finite limit, say $I_\infty(Y)$, with $E[I_\infty(Y)] = 0$ and

$$E[I_\infty^2(Y)] = E\left[\int_0^T X^2(s)\,ds\right] < \infty.$$

To conclude the proof observe that $I_\infty(Y) = Z(T)$.

Problem 4.2 In the setting of Section 4.11, suppose that f is a C^2 function that satisfies $\lambda f - \Gamma f = 0$ on $[a, b]$ with $f(a) = 1$ and $f(b) = 0$. Show that $f(x) = E_x\{\exp(-\lambda T); X_T = a\}$. When combined with Problem 3.3, this verifies the formula for $\psi_1(x)$ derived in Section 3.3, and the formula for $\psi_2(x)$ can be verified in the same way.

Problem 4.3 In the setting of Section 4.11, let $\psi_1(x) := E_x\{\exp(-\lambda T); X_T = a\}$ and $\psi_2(x) := E_x\{\exp(-\lambda T); X_T = b\}$. (This generalizes slightly the notation of Section 3.3, which was restricted to the case $a = 0$.) Show that

$$E_x\left[e^{-\lambda T}f(X_T)\right] = f(a)\psi_1(x) + f(b)\psi_2(x).$$

If $|f|$ is bounded by a polynomial, then the right side goes to zero as $a \to -\infty$, $b \to \infty$. Prove this statement, using the formulas for ψ_1 and ψ_2 developed in Section 3.3.

Problem 4.4 (*Continuation*) Let $u : \mathbb{R} \to \mathbb{R}$ be continuous with $|u|$ bounded by a polynomial. Suppose that f satisfies

(4.62) $$\lambda f - \Gamma f = u$$

on $[a, b]$ and $f(a) = f(b) = 0$. Show that

$$f(x) = E_x\left[\int_0^T e^{-\lambda t}u(X_t)\,dt\right].$$

Next, dropping the requirement that $f(a) = f(b) = 0$, suppose that (4.62) holds on all of $\mathbb{R}$ and that $|f|$ is bounded by a polynomial. Show that

$$f(x) = E_x\left[\int_0^\infty e^{-\lambda t}u(X_t)\,dt\right].$$

Problem 4.5 Altering slightly the setup in Section 4.11, suppose that f is continuous, that f is twice continuously differentiable except at an isolated point y ($a < y < b$), that $\Gamma f(x) = 0$ except at $x = y$, that $\sigma^2 \Delta f'(y) = -2$ (see Section 3.4 for the exact meaning of this condition), and that $f(a) = f(b) = 0$. Use the generalized Itô formula (4.47) to show that

$$f(x) = E_x\left[l(T, y)\right].$$

Problem 4.6 (*Continuation*)   With $\lambda > 0$, now suppose that $\lambda f - \Gamma f = 0$ except at $x = y$. All other assumptions are as before. Show that

$$f(x) = E_x \left[\int_0^T e^{-\lambda t} l\,(dt, y) \right].$$

This requires that (4.47) be combined with the specialized integration by parts formula of Proposition 4.15; the structure of the argument is the same as that used to prove Proposition 4.17.

Problem 4.7   Use Itô's formula to explicitly calculate $\int W^9\,dW$. Any expression not involving a stochastic integral is considered an answer.

Problem 4.8   In the setting of Section 4.11, suppose that f is C^2 on $[0, b]$ with

(4.63) $\Gamma f = -1$ on $[0, b]$ and $f(0) = f(b) = 0.$

Show that $f(x) = E_x(T)$. Show that the expression for $E_x(T)$ developed in Problem 3.12 does in fact satisfy (4.63).

Problem 4.9   In the setting of Section 4.11, let $f_n(x) := E_x[\int_0^T X_t^n\,dt]$. Use Itô's formula to develop a general formula for $f_n(x)$.

Problem 4.10   Let $f(t, x)$ be twice continuously differentiable on $\mathbb{R}^2$. Let μ and $\sigma > 0$ be constants and define

$$g(t, x) := \frac{\partial}{\partial x} f(t, x)$$

and

$$h(t, x) := \left(\frac{\partial}{\partial t} + \frac{1}{2}\sigma^2 \frac{\partial^2}{\partial x^2} + \mu \frac{\partial}{\partial x} \right) f(t, x).$$

Adopting the setup of Section 4.11, let $T < \infty$ be a stopping time. Use the multi-dimensional Itô formula (4.38) to show that

$$f(T, X_T) = f(0, X_0) + \sigma \int_0^T g(t, X_t)\,dW + \int_0^T h(t, X_t)\,dt.$$

Problem 4.11 (*Continuation*)   Fix $t > 0$, assume that $X_0 > 0$, and set $T := T(0) \wedge t$. Suppose that $g(s, y)$ is bounded on $[0, t] \times [0, \infty)$ and that $h(s, y) = 0$ on $[0, t] \times [0, \infty)$. Use Corollary 4.8 to show that

$$f(0, x) = E_x\left[f(T, X_T) \right].$$

Finally, assume that $f(s, 0) = 0$ for $0 \leq s \leq t$ and conclude that

$$f(0, x) = E_x[f(t, X_t); \, T(0) > t].$$

Problem 4.12 (*Continuation*)   Suppose that $G(t, x)$ is defined on $[0, \infty) \times [0, \infty)$ and is twice continuously differentiable up to the boundary. That is, all first- and second-order partials approach finite limits at all boundary points and those limits are continuous functions on the boundary; this condition assures that G can be extended to a function twice continuously differentiable on all of $\mathbb{R}^2$. Let $u : \mathbb{R} \to \mathbb{R}$ be bounded and continuous, and suppose that G satisfies

(a) $\qquad \dfrac{\partial}{\partial t} G(t, x) = \left(\dfrac{1}{2}\sigma^2 \dfrac{\partial^2}{\partial x^2} + \mu \dfrac{\partial}{\partial x} \right) G(t, x) \qquad$ for $t, x \geq 0$

(b) $\qquad G(t, 0) = 0 \qquad\qquad\qquad\qquad\qquad$ for $\;\; t \geq 0$

(c) $\qquad G(0, x) = u(x) \qquad\qquad\qquad\qquad\;\;$ for $\;\; x \geq 0$

(d) $\qquad \dfrac{\partial}{\partial x} G(t, x)$ is bounded on $[0, \infty) \times [0, \infty)$.

Now fix $t > 0$ and define $f(s, x) = G(t - s, x)$ for $0 \leq s \leq t$ and $x \geq 0$, observing that f can be extended to a function on $\mathbb{R}^2$ satisfying all the conditions of Problem 4.11. Use the result of that problem to conclude that

$$G(t, x) = E_x[u(X_t); \, T(0) > t].$$

Problem 4.13 (*Continuation*)   Fix $y \geq 0$ and let $G(t, x, y)$ be defined by formula (3.48). Viewed as a function of t and x alone, this particular G does *not* satisfy the assumptions of Problem 4.12 because of discontinuities at $t = 0$. But for any $\epsilon > 0$ the function $G^*(t, x) := G(t + \epsilon, x, y)$ does satisfy all the stated conditions (see Problem 3.10); apply the result of Problem 4.12 to conclude that

$$G^*(t, x) = E_x[G^*(0, X_t); \, T(0) > t]$$

or equivalently

$$G(t + \epsilon, x, y) = E_x[G(\epsilon, X_t, y); \, T(0) > t].$$

Recalling that $G(0, x, y) = 1_{(x>y)}$, let $\epsilon \downarrow 0$ and use the bounded convergence theorem to conclude that

$$G(t, x, y) = P_x\{X_t > y; \, T(0) > t\}.$$

This verifies the interpretation for G given in Section 3.5.

Problem 4.14 In the setting of Section 4.11, let $T := \inf\{t \ge 0 : |X_t| \ge \epsilon\}$, where $\epsilon > 0$. Apply Proposition 4.17 with $\lambda = 0$ and $f(x) = x^2$ to conclude that $\sigma^2 E_0(T) \sim \epsilon^2$ as $\epsilon \downarrow 0$.

Problem 4.15 Consider the situation hypothesized in Proposition 4.17, but with the function $f : \mathbb{R} \to \mathbb{R}$ modified in the following way: it is twice continuously differentiable *except* at the origin, where $\Delta f'(0) = \delta > 0$. Using the generalized Itô formula (4.47), modify the proof of Proposition 4.17 to obtain the following identity:

$$f(0) = E_0\left\{ \int_0^T e^{-\lambda t}\left[u(X_t)\, dt - \tfrac{1}{2}\delta\sigma^2 l(dt, 0) \right] \right\} + E_0\left[e^{-\lambda T} f(X_T) \right],$$

where $u := \lambda f - \Gamma f$ away from the origin, as in Proposition 4.17.

Problem 4.16 (*Continuation*) Given $\epsilon > 0$, let $a = -\epsilon$ and $b = \epsilon$, so that $T = \inf\{t \ge 0 : |X_t| \ge \epsilon\}$. Specializing the result in Problem 4.15 to this stopping time and the function $f(x) := |x|$, and using the result of Problem 4.14, show that

$$E_0\left[\int_0^T e^{-\lambda t} l(dt, 0) \right] \sim \frac{1}{\sigma^2}\epsilon \qquad \text{as } \epsilon \downarrow 0.$$

With $\lambda = 0$ this reduces to $\sigma^2 E_0[l(T, 0)] \sim \epsilon$ as $\epsilon \downarrow 0$.

Problem 4.17 (*Continuation*) Let f be as in Problem 4.15 and T as in Problem 4.16. Use the results of Problems 4.14 to 4.16 to conclude the following:

$$E_0\left[e^{-\lambda T} f(X_T) \right] > f(0) \qquad \text{for all } \epsilon > 0 \text{ sufficiently small.}$$

Problem 4.18 Consider a one-dimensional diffusion process X (refer to Section 4.5) whose coefficient functions $\mu(\cdot)$ and $\sigma(\cdot)$ are continuous and hence bounded on finite intervals. Show that Propositions 4.16 and 4.17 remain valid for this process X with the differential operator Γ redefined as follows:

$$\Gamma f(x) := \mu(x)f'(x) + \tfrac{1}{2}\sigma^2(x)f''(x), \qquad x \in \mathbb{R}.$$

5

Optimal Stopping of Brownian Motion

In the general formulation of an optimal stopping problem, the basic elements are an $\mathbb{R}^n$-valued stochastic process $X = \{X_t, \ t \geq 0\}$ that is adapted to a given filtration, and a measurable payoff function $g : \mathbb{R}^n \to \mathbb{R}$. The decision maker's problem is then to choose a stopping time T to

$$(5.1) \qquad \text{maximize } E\left[e^{-\lambda T}g(X_T)\right],$$

where $\lambda \geq 0$ is the interest rate for discounting, also given. Attention will be restricted here to problems where X is one-dimensional, but the following observation is important for understanding the more general literature: if X is allowed to be multi-dimensional, inclusion of the discount factor $\exp(-\lambda T)$ in (5.1) is superfluous, because one can add the deterministic process $\exp(-\lambda t)$ as an extra component of X_t and then incorporate discounting in the definition of g.

The mathematical theory of optimal stopping is highly developed, featuring incisive theorems that allow very general processes X and very general payoff functions g. Furthermore, for the special case where X is a one-dimensional diffusion process, the optimal stopping problem has been completely solved in a certain sense. The general solution will be described in Section 5.5 for the even more special case where X is a Brownian motion, and readers will see that it provides useful qualitative insights, but it is seldom applied in the analysis of specific problems. Rather, problems that arise in applications are typically solved from first principles, using stochastic calculus and the "principle of smooth fit" in a guess-and-verify process to be illustrated in this chapter.

Two specific problems will be formulated and solved in the sections that follow, including the influential investment model originally propounded by McDonald and Siegel (1986). The process X underlying these problems is a one-dimensional Brownian motion, and in each case the payoff function g has a particular functional form, but the approach that we use is applicable in much more general settings.

5.1 A general problem

Given parameters μ and $\sigma > 0$, let $(\Omega, \mathcal{F}, \mathbb{F}, P_x)$ be the canonical space defined in Section 3.1, and let X be the coordinate process on Ω. Thus X is a (μ, σ) Brownian motion with starting state x on $(\Omega, \mathcal{F}, \mathbb{F}, P_x)$. Let there be given a continuous function $g : \mathbb{R} \to [0, \infty)$ and an interest rate $\lambda \geq 0$ for discounting. For an arbitrary stopping time T we define the associated *value function* $v(\cdot)$ as follows:

$$(5.2) \qquad v(x) := E_x\left[e^{-\lambda T}g(X_T)\right], \qquad x \in \mathbb{R},$$

with the random variable in square brackets defined to be zero on the set $\{T = \infty\}$. Because $g(\cdot)$ is positive by assumption, the expectation in (5.2) is well defined, but it may be $+\infty$. A stopping time T is called *optimal* if its value function majorizes that for any other stopping time.

To ensure that $v(\cdot)$ is well defined for an arbitrary stopping time T, we have assumed that $g(\cdot)$ is positive, but that assumption is not actually restrictive, for the following reason. If g *were* allowed to take negative values, a rational decision maker would never stop in a state x where $g(x) < 0$, because the decision maker may choose to never stop (that is, may choose $T = \infty$), and that choice provides zero payoff as a matter of definition. Thus the decision maker's optimization problem is essentially unchanged if we replace $g(\cdot)$ by $g^+(\cdot)$.

Theorem 5.1 *Suppose that $\varphi : \mathbb{R} \to \mathbb{R}$ is C^1 and piecewise C^2 (see Guide to Notation and Terminology for the precise meaning of that term), and further satisfies the following:*

$$(5.3) \qquad\qquad \varphi(x) \geq g(x) \qquad \text{for all } x \in \mathbb{R};$$
$$(5.4) \qquad \lambda\varphi(x) - \Gamma\varphi(x) \geq 0 \qquad \text{for almost all } x \in \mathbb{R},$$

where $\Gamma\varphi(x) = \frac{1}{2}\sigma^2\varphi''(x) + \mu\varphi'(x)$ as in Chapters 3 and 4. Then φ majorizes the value function for every stopping time.

Corollary 5.2 *If the value function $v(\cdot)$ for a stopping time T satisfies all the hypotheses of Theorem 5.1, then T is optimal.*

Proof Let T be an arbitrary stopping time, let $T(b)$ be the first time $t \geq 0$ at which $|X_t| \geq b > 0$, and define $\tau(b) := T \wedge T(b)$. Also, let $u := \lambda\varphi - \Gamma\varphi$, with $u(x)$ defined arbitrarily at points x where φ'' is discontinuous. Applying Proposition 4.17 with φ in place of f, we have that

$$(5.5) \qquad \varphi(x) = E_x\left[\int_0^{\tau(b)} e^{-\lambda t}u(X_t)\,dt\right] + E_x\left[e^{-\lambda\tau(b)}\varphi(X_{\tau(b)})\right].$$

Thus (5.3) and (5.4) give

$$(5.6) \qquad \varphi(x) \geq E_x \left[e^{-\lambda \tau(b)} g(X_{\tau(b)}) \right].$$

Because g is positive, it is immediate from (5.6) that

$$(5.7) \qquad \varphi(x) \geq E_x \left[e^{-\lambda T} g(X_T); \ T \leq T(b) \right].$$

The parameter b was chosen arbitrarily and $T(b) \to \infty$ almost surely as $b \to \infty$, so upon letting $b \to \infty$ in (5.7) we have

$$(5.8) \qquad \varphi(x) \geq E_x \left[e^{-\lambda T} g(X_T) \right].$$

This proves Theorem 5.1, and the corollary is then immediate. $\qquad\square$

Corollary 5.2 provides the means to rigorously verify that a particular stopping time, perhaps derived by a nonrigorous argument in the context of a concrete problem, is in fact optimal. Such results are often referred to as *verification theorems*. The guess-and-verify process will be illustrated in Sections 5.3 and 5.5.

5.2 Continuation costs

Maintaining the general setup of the previous section, let us now assume that $\lambda > 0$ (that is, the interest rate for discounting is *strictly* positive). Given a continuous cost function $u : \mathbb{R} \to [0, \infty)$ that satisfies

$$(5.9) \qquad f(x) := E_x \left[\int_0^\infty e^{-\lambda t} u(X_t) \, dt \right] < \infty \qquad \text{for all } x \in \mathbb{R},$$

consider the following, apparently more general problem: choose a stopping time T to

$$(5.10) \qquad \text{maximize } E_x \left[e^{-\lambda T} g(X_T) \right] - E_x \left[\int_0^T e^{-\lambda t} u(X_t) \, dt \right].$$

That is, in addition to the terminal reward $g(X_T)$ that is received at the time of stopping, costs are continuously incurred at rate $u(X_t)$ up until the time of stopping.

This problem actually reduces to the one stated in Section 5.1, but with a different payoff function $\hat{g}$, as follows. First, using the strong Markov property of X exactly as in the derivation leading up to Proposition 3.8, we have

$$(5.11) \qquad E_x \left[\int_0^T e^{-\lambda t} u(X_t) \, dt \right] = f(x) - E_x \left[e^{-\lambda T} f(X_T) \right].$$

Of course, $f(x)$ does not depend on the chosen stopping time T, so the objective (5.10) is equivalent to the following:

$$(5.12) \qquad \text{maximize } E_x\left[e^{-\lambda T}\hat{g}(X_T)\right],$$

where $\hat{g} = g + f$. Note that f is positive, continuous, and finite-valued by assumption, so $\hat{g}$ does in fact satisfy the assumptions stated in Section 5.1.

5.3 McDonald–Siegel investment model

Specializing the stopping problem in Section 5.1, we now assume that $\lambda > 0$, that

$$(5.13) \qquad\qquad \gamma := \mu + \tfrac{1}{2}\sigma^2 < \lambda,$$

and that

$$(5.14) \qquad g(x) = (ke^x - I)^+, \quad x \in \mathbb{R}, \quad \text{where } k, I > 0.$$

To motivate this very specific problem, we define the positive process $Y_t := \exp(X_t)$, $t \geq 0$, so the objective (5.1) can be restated as follows:

$$(5.15) \qquad\qquad \text{maximize } E\left[e^{-\lambda T}(kY_T - I)^+\right].$$

Obviously, if $X_0 = x \in \mathbb{R}$ then the initial value of Y is $Y_0 = \exp(x)$. Representing X as $X_t = X_0 + \mu t + \sigma W_t$, where W is a standard Brownian motion, it follows from Itô's formula that Y satisfies the stochastic differential equation

$$(5.16) \qquad\qquad dY = \gamma Y\,dt + \sigma Y\,dW.$$

The process Y is called *geometric Brownian motion* (or less commonly, *exponential Brownian motion*), and it is widely used in financial economics as a stochastic model of price dynamics. For example, Y_t might be interpreted as the price commanded by q tons of copper on world markets at time t. Now consider an investment opportunity that provides a perpetual flow of q tons of copper per time unit to the investor, in exchange for an initial investment of I dollars. Recall from Section 1.4 that

$$E(e^{X_t}) = e^{X_0}e^{(\mu + \sigma^2/2)t} = Y_0 e^{\gamma t}.$$

Thus, if the investment is made at time zero, the expected present value of the copper it provides is

$$E\left(\int_0^\infty e^{-\lambda t}Y_t\,dt\right) = \int_0^\infty e^{-\lambda t}Y_0 e^{\gamma t}\,dt = kY_0,$$

where $k = (\lambda - \gamma)^{-1}$, and hence the expected net present value of the investment is $kY_0 - I$. More generally, if the investment is postponed until a future time T, its expected net present value at the moment of investment is $kY_T - I$. As noted immediately before Theorem 5.1, a rational decision maker will never invest at a time when that figure is negative, so the investor's optimization problem can be expressed in the form (5.15). Of course, the McDonald–Siegel stopping problem (5.15) is less specific than the story just told: it simply assumes that the expected net present value of the cash flows generated by the investment, starting from future times $t \geq 0$, can be modeled as a geometric Brownian motion.

A reasonable guess is that there exists an optimal policy of the following form: stop at the first time t when Y_t exceeds some *trigger level* y^*; that is, choose a stopping time of the form

$$(5.17) \qquad\qquad T = \inf\{t \geq 0 : X_t \geq b\},$$

where $b = \log y^*$. Our problem now is to choose the barrier height b and to rigorously verify that the corresponding stopping time T is optimal. For $X_0 = x < b$, we know from Section 3.3 that $E_x[\exp(-rT)] = \exp(-\alpha_2(b - x))$ where $\alpha_2 := \alpha_2(\lambda)$ is given by formula (3.23). Thus the return function $v(\cdot)$ corresponding to T is

$$(5.18) \qquad v(x) = \begin{cases} e^{-\alpha_2(b-x)}g(b) & \text{if } x < b \\ g(x) & \text{if } x \geq b. \end{cases}$$

Before proceeding further, we record the following important consequence of assumption (5.13).

Proposition 5.3 $\alpha_2 > 1.$

Proof From the definition (3.23) of α_2 one sees that $\alpha_2 > 1$ if and only if

$$(5.19) \qquad (\mu^2 + 2\lambda\sigma^2)^{1/2} > \mu + \sigma^2 = \gamma + \tfrac{1}{2}\sigma^2.$$

If $\gamma + \tfrac{1}{2}\sigma^2 \leq 0$, the desired conclusion is then immediate. Otherwise, we can square both sides of (5.19), make the substitution $\mu = \gamma - \tfrac{1}{2}\sigma^2$, and rearrange terms to conclude the following: $\alpha_2 > 1$ if and only if $2\lambda\sigma^2 > 2\gamma\sigma^2$. But assumption (5.13) is that $\gamma < \lambda$, so the proposition is proved. $\square$

From (5.18) we have that $v'(b+) = ke^b$ and $v'(b-) = \alpha_2 g(b)$, so $v'(\cdot)$ is continuous at b if and only if b is chosen to satisfy

$$(5.20) \qquad ke^b = \left(\frac{\alpha_2}{\alpha_2 - 1} \right) I.$$

Proposition 5.4   *If b is chosen to satisfy* (5.20), *then the value function v defined by* (5.18) *satisfies all the hypotheses of Theorem 5.1, and hence the stopping time T defined by* (5.17) *is optimal.*

Proof   Substituting (5.20) into (5.14) we have that

$$(5.21) \qquad\qquad g(b) = (\alpha_2 - 1)^{-1} I.$$

By construction, the value function $v(x)$ is continuously differentiable, and its second derivative exists and is continuous except possibly at $x = b$. To show that v majorizes g, we first use (5.18) to write

$$(5.22) \qquad\qquad v(b - y) = g(b)e^{-\alpha_2 y}, \qquad y \geq 0.$$

On the other hand, for $y \geq 0$ small enough to ensure $g(b - y) > 0$ we have

$$(5.23) \qquad\qquad g(b - y) = ke^{b-y} - I.$$

Let $\Delta(y) := (\alpha_2 - 1)[v(b - y) - g(b - y)]/I$ for $y \geq 0$. Substracting (5.23) from (5.22), using (5.21) and simplifying, we arrive at the following:

$$(5.24) \qquad\qquad \Delta(y) = e^{-\alpha_2 y} - \alpha_2 e^{-y} + \alpha_2 - 1, \qquad g \geq 0.$$

From this we see that $\Delta(y) > 0$ for all $y > 0$, because $\Delta(0) = \Delta'(0) = 0$ but $\Delta''(y) > 0$ for all $y > 0$. Thus $v(x) > g(x)$ for all $x < b$. Of course, $v(x) = g(x)$ for $x \geq b$, so v majorizes g.

Finally, we need to show that $\Gamma v(x) \leq \lambda v(x)$ for all $x \in \mathbb{R}$. Because the exponential function $\theta(x) := \exp(\alpha_2 x)$ satisfies $\Gamma\theta = \lambda\theta$ (see Section 3.3), we have from (5.18) that $\Gamma v(x) = \lambda v(x)$ for $x < b$. On the other hand, $v(x) = g(x)$ for $x \geq b$, so it remains to show that $\Gamma g(x) \leq \lambda g(x)$ for $x \geq b$. By construction we have that $g(b+) = v(b-)$ and $g'(b+) = v'(b-)$. Also, readers can easily verify that $v''(b-) = \alpha_2 g''(b+) > 0$. Because $\alpha_2 > 1$ we then have

$$(5.25) \qquad\qquad (\Gamma - \lambda)g(b+) < (\Gamma - \lambda)v(b-) = 0.$$

Because $g(b + y) = ke^{b+y} - I$ for $y \geq 0$, we have

$$(5.26) \qquad (\Gamma - \lambda)g(b + y) = (\Gamma - \lambda)g(b+)e^y - \lambda(e^y - 1)I$$

for all $y \geq 0$. Combining (5.25) and (5.26) gives $(\Gamma - \lambda)g(b + y) < 0$ for all $y \geq 0$, which completes the proof.   □

In this analysis we have chosen the policy parameter b so as to ensure that the value function $v(\cdot)$ is continuously differentiable at b, and then proved that the associated stopping time is optimal. This criterion for policy design is known as the *principle of smooth fit*, and as we shall see in

Section 5.5, it is broadly applicable to optimal stopping problems where the underlying process X is one-dimensional. That discussion is complemented by Problems 5.1 and 5.2, where readers are asked to show by direct means that continuity of $v'(\cdot)$ at b is a *necessary* condition for optimality in the McDonald–Siegel investment model.

An alternative, more direct way to determine the optimal value of the policy parameter b is to differentiate with respect to b in (5.18) for arbitrary $x < b$, and set the derivative to zero. This produces the same optimality criterion (5.20), irrespective of the x value with which one starts, provided that x is sufficiently small. Perhaps surprisingly, however, the mechanics of solution are typically simpler with the smooth-fit method than with the direct approach.

5.4 An investment problem with costly waiting

Suppose that the problem posed in the previous section is changed in the following two ways. First, to simplify calculations, let the payoff function upon stopping be

$$(5.27) \qquad g(x) = (x - I)^{+}, \qquad x \in \mathbb{R},$$

which means that the value of the investment is modeled by the (μ, σ) Brownian motion X itself, rather than the geometric Brownian motion $Y = k \exp(X)$. Second, suppose that the decision maker continuously incurs cost at rate $c > 0$ up until either (a) the investment is made, or (b) the option to invest is irreversibly relinquished. That is, we now consider a model variant in which the decision maker must pay c dollars per time unit to maintain the right to invest later. This is a special case of the situation described in Section 5.2, with continuation cost function $u(x) = c$, and hence the decision maker's original problem reduces to the optimal stopping problem of Section 5.1 with the following alternative payoff function:

$$(5.28) \qquad \begin{aligned} \hat{g}(x) &:= c/\lambda + g(x) = c/\lambda + (x - I)^{+} \\ &= c_1 \vee (x + c_2), \qquad x \in \mathbb{R}, \end{aligned}$$

where

$$(5.29) \qquad c_1 := c/\lambda > 0 \quad \text{and} \quad c_2 := c/\lambda - I.$$

For an intuitive understanding of this equivalent cost structure one may imagine the following: at time zero the decision maker pays out c/λ dollars, which is the cost of maintaining forever the right to invest without ever actually exercising that right; and then c/λ dollars are paid back to the decision maker at the stopping time T that he or she chooses. The payment at time zero does not depend on the choice of T, so it can be ignored, and hence we arrive at the equivalent payoff function $\hat{g}$ in (5.28).

With the cost structure modified in this way, a plausible guess is that there exists an optimal stopping time of the following form:

$$(5.30) \qquad T = \inf\{t \geq 0 : X_t \leq a \text{ or } X_t \geq b\},$$

where $-\infty < a < b < \infty$, $g(a) = c_1$, and $g(b) = c_2 + x$. From the development in Section 3.6 we know that the associated value function v must satisfy

$$(5.31) \qquad \lambda v(x) - \Gamma v(x) = 0, \qquad a \leq x \leq b,$$

with boundary conditions

$$(5.32) \qquad v(a) = \hat{g}(a) = c_1 \quad \text{and} \quad v(b) = \hat{g}(b) = c_2 + b,$$

and the principle of smooth fit (see Section 5.3) suggests that optimal trigger values a and b can be determined by the following conditions:

$$(5.33) \qquad v'(a+) = \hat{g}'(a-) = 0,$$
$$(5.34) \qquad v'(b-) = \hat{g}'(b+) = 1.$$

To construct a solution of (5.31) to (5.34) we start by defining

$$(5.35) \qquad f(y) := (\alpha_1 + \alpha_2)^{-1}(\alpha_2 e^{-\alpha_1 y} + \alpha_1 e^{\alpha_2 y}), \qquad y \geq 0,$$

where $\alpha_1 := \alpha_1(\lambda)$ and $\alpha_2 := \alpha_2(\lambda)$ are defined as in Section 3.3. This function f is the unique solution of the differential equation $\lambda f - \Gamma f = 0$ subject to boundary conditions $f(0) = 1$ and $f'(0) = 0$. Thus by defining

$$(5.36) \qquad v(x) = c_1 f(x - a), \qquad a \leq x \leq b,$$

we will satisfy (5.31), $v(a) = c_1$ and $v'(a+) = 0$, regardless of how a and b are chosen. Defining $\delta := b - a > 0$, we have from (5.34) and (5.36) that $c_1 f'(\delta) = 1$, or equivalently,

$$(5.37) \qquad f'(\delta) = \lambda/c.$$

It is easy to verify that $f(\cdot)$ is convex and strictly increasing on $[0, \infty)$, so there is a unique $\delta > 0$ that satisfies (5.37). Finally, combining (5.36) with the requirement $v(b) = c_2 + b$ in (5.32) gives the following:

$$(5.38) \qquad c_1 f(\delta) = c_2 + b.$$

Proposition 5.5 *Let $\delta > 0$ and $b \in \mathbb{R}$ be determined by (5.37) and (5.38), and define $a := b - \delta$. Also, let $v : \mathbb{R} \to \mathbb{R}$ be defined by (5.36) on $[a, b]$, with $v(x) = c_1$ for $x < a$ and $v(x) = x + c_2$ for $x > b$. Then v is the value function for the stopping time T defined by (5.30), and it satisfies all the hypotheses of Theorem 5.1.*

Corollary 5.6 *T is optimal.*

Proof Because f is strictly increasing and strictly convex on $[0, \infty)$, it is apparent that the constructed function v majorizes the piecewise linear payoff function $\hat{g}$ on $[a, b]$, and it coincides with $\hat{g}$ outside that interval. The first part of the conclusion is then immediate from the constructed properties (5.31) and (5.32). To establish that v satisfies all the hypotheses of Theorem 5.1 it remains only to show that $\lambda \hat{g}(x) \geq \Gamma \hat{g}(x)$ outside the interval $[a, b]$. That inequality is trivial for $x \leq a$, because $\hat{g}(x) = c_1 > 0$ and $\Gamma \hat{g}(x) = 0$ in that range. Also, for $x \geq b$ we have $\Gamma \hat{g} \equiv \mu$, so $\lambda \hat{g} - \Gamma \hat{g}$ is strictly increasing in that range, and it remains only to show that

$$(5.39) \qquad (\lambda - \Gamma)\hat{g}(b+) := \lambda \hat{g}(b) - \mu \geq 0.$$

From (5.31) we know that $(\lambda - \Gamma)v(b-) = 0$, and from the boundary conditions (5.32) and (5.34) it follows that

$$(\lambda - \Gamma)\hat{g}(b+) - (\lambda - \Gamma)v(b-) = \tfrac{1}{2}\sigma^2 \left[f''(\delta) - \hat{g}''(b) \right].$$

But $f''(\delta) > 0$ and $\hat{g}''(b) = 0$, so (5.39) is established. Finally, Corollary 5.6 is an application of Corollary 5.2. $\qquad\square$

5.5 Some general theory

Returning to the general problem laid out in Section 5.1, we shall state in this section several important results that are either special cases of theorems proved by Dayanik and Karatzas (2003) or else follow directly from one of their theorems. These results will not be proved, but their intuitive basis will be explained. At the end of the section there is a brief and informal discussion of a more general problem: optimal stopping of a one-dimensional diffusion process.

A central object of study in the general theory of optimal stopping is the *optimal value function*, which is defined as follows in the setting of Section 5.1:

$$(5.40) \qquad V(x) := \sup E_x \left[e^{-\lambda T} g(X_T) \right], \qquad x \in \mathbb{R},$$

where the supremum is taken over all stopping times T. The following proposition gives necessary and sufficient conditions for V to be finite-valued. Here $\alpha_1 := \alpha_1(\lambda)$ and $\alpha_2 := \alpha_2(\lambda)$ are defined as in Section 3.3 for $\lambda > 0$, and those definitions are extended in the obvious way to the case $\lambda = 0$, namely, $\alpha_1(0) := 0$ and $\alpha_2(0) := 2|\mu|/\sigma^2$ when $\mu \leq 0$, whereas $\alpha_1(0) := 2\mu/\sigma^2$ and $\alpha_2(0) := 0$ when $\mu \geq 0$.

Proposition 5.7 $V(x) < \infty$ *for all* $x \in \mathbb{R}$ *if and only if the following conditions are satisfied:*

$$(5.41) \qquad\qquad \sup\{e^{\alpha_1 y}g(y) : y \leq 0\} < \infty$$

and

$$(5.42) \qquad\qquad \sup\{e^{-\alpha_2 y}g(y) : y \geq 0\} < \infty.$$

The necessity of (5.41) for finiteness of V is more or less obvious, as follows. Defining $T(y) := \inf\{t \geq 0 : X_t \leq y\}$, we know from Section 3.3 that the associated value function is $v(x) = \exp\{-\alpha_1(x - y)g(y)\}$ for $x \geq y$, and if (5.41) did *not* hold we could choose y sufficiently negative to make $v(x)$ arbitrarily large, implying that $V(x) = \infty$. There is a symmetric argument for the necessity of (5.42), but proving the sufficiency of these conditions requires more effort. To avoid distractions we assume hereafter that (5.41) and (5.42) hold.

Loosely adapting a standard term in Markov process theory, we say that a function $f : \mathbb{R} \to \mathbb{R}$ is λ-excessive for X, or simply *excessive*, if it is C^2 and satisfies the differential inequality (5.4) on all of $\mathbb{R}$. Also, we denote by $\mathcal{E}(g)$ the class of excessive functions that majorize g, that is, the set of all excessive functions f such that $f(x) \geq g(x)$, $x \in \mathbb{R}$. If v is the value function for a stopping time T and $f \in \mathcal{E}(g)$, then $f \geq v$ by Theorem 5.1, so obviously $f \geq V$ as well.

Our next proposition provides an elegant characterization of the optimal value function V. It does not appear directly in the paper by Dayanik and Karatzas (2003), but it can easily be deduced from their results and the following lemma: any function of the type those authors call F-concave can be uniformly approximated on an arbitrary compact interval by another F-concave function which is C^2; that lemma can in turn be deduced from the more familiar fact that any concave function on a compact interval can be uniformly approximated by another concave function that is C^2.

Proposition 5.8 $V(x) = \inf\{f(x) : f \in \mathcal{E}(g)\}$, $x \in \mathbb{R}$.

To repeat an earlier statement, it is clear from Theorem 5.1 that the infimum in Proposition 5.8 provides an upper bound on V, and the following argument makes it plausible that the two actually coincide. Suppose for the moment that the optimal value function V, which is defined by (5.40), is itself C^2. It is then quite easy to show that V satisfies the differential inequality (5.4); see Problem 5.3. Of course, V must also majorize the payoff function g, which means that V is itself an element of $\mathcal{E}(g)$, which proves Proposition 5.8. The flaw in this argument is that V need not be C^2 (in fact, it need not be continuously differentiable if g is not), but Proposition 5.8 is true nonetheless, and from it the following are obtained.

Proposition 5.9 *If g is continuously differentiable at a point $x \in \mathbb{R}$ where $V(x) = g(x)$, then V is continuously differentiable at x as well.*

Proposition 5.10 *The stopping time $T := \{\inf t \geq 0 : V(X_t) = g(X_t)\}$ is optimal.*

Informed by Proposition 5.10, we define $S := \{x \in \mathbb{R} : V(x) = g(x)\}$, calling S the optimal *stopping region* and calling its complement the optimal *continuation region*. Proposition 5.9 states precisely the "principle of smooth fit" that was invoked in Sections 5.3 and 5.4. In each of those settings there is a point $x_0 \in \mathbb{R}$ where the derivative of the payoff function (g in the former case, $\hat{g}$ in the latter case) is discontinuous, but we guessed (and later confirmed) that x_0 would not appear as a boundary point of the optimal stopping region, implying by Proposition 5.10 that $V'(\cdot)$ is continuously differentiable at the boundary of the stopping region. There do exist optimal stopping problems where the optimal value function has a discontinuous derivative (see Problems 5.5 and 5.6), but they seem to be rare in well-motivated applications.

To conclude this section we note that the general theory outlined above goes through with very little change when X is a diffusion process (see Section 4.5), satisfying a stochastic differential equation of the form

$$(5.43) \qquad dX = \mu(X)\,dt + \sigma(X)\,dW,$$

where W is a standard Brownian motion and $\mu(\cdot)$ and $\sigma(\cdot)$ are sufficiently regular functions. One good example is the geometric Brownian motion that was introduced in Section 5.3, and other examples will appear in Chapters 8 and 9. In this more general setting one must redefine the differential operator Γ as in Problem 4.18, that is,

$$\Gamma f(x) := \mu(x)f'(x) + \tfrac{1}{2}\sigma^2(x)f''(x),$$

after which all other definitions and results remain valid as stated, except for Proposition 5.8, which must be expressed in substantially altered terms.

5.6 Sources and literature

In addition to the seminal paper by McDonald and Siegel (1986), Dixit and Pindyck (1994) provide a thorough but accessible treatment of the investment model described in Section 5.3, and of various related models that arise naturally in economics and finance. The general theory recounted in Section 5.5 is drawn from the paper by Dayanik and Karatzas (2003), who restrict attention not only to time-homogeneous, one-dimensional diffusion processes, but also to time-homogeneous payoff functions. That is, Dayanik and Karatzas do *not* consider problems where the payoff depends on both the state and the time at which stopping occurs. (Some of the optimal stopping problems of greatest interest in mathematical finance fall in the excluded class, because they involve an option that is worthless after a given expiration date.) That restricted scope leads both to razor-sharp results and a clean, elegant mathematical development.

Øksendal (2007) provides an excellent account of optimal stopping theory for multi-dimensional diffusions, including problems where payoff depends on both state and time; see also Appendix D of Karatzas and Shreve (1991) and the references therein. Peskir and Shiryaev (2006) provide a still more general theoretical development, not restricted to diffusion processes, and a thorough treatment of specially structured stopping problems that arise in finance and statistics. Chapter 8 of this book gives an introduction to the statistical models that they consider.

5.7 Problems and complements

Problem 5.1   In the McDonald–Siegel investment model (Section 5.3), show that if the barrier height b is chosen so that $v'(b-) > g'(b+)$, then $v(x) < g(x)$ for $x < b$ sufficiently close to b, and hence the associated stopping time is not optimal.

Problem 5.2 (*Continuation*)   Suppose, on the other hand, that b is chosen so that $v'(b-) < g'(b+)$. Using the result of Problem 4.17, show that there exists a stopping time T such that $E_b[\exp(-\lambda T)v(X_T)] > v(b)$. Conclude that the stopping time associated with b is not optimal.

Problem 5.3   Suppose that f is C^2 and that $\Gamma f(x) > \lambda f(x)$ for some $x \in \mathbb{R}$.

Use Proposition 4.17 to show that for $\epsilon > 0$ sufficiently small and $T :=$ $\inf\{t \geq 0 : |X_t - x| \geq \epsilon\}$ one has

$$E_x\left[e^{-\lambda T}f(X_T)\right] > f(x).$$

Use this to show that if the optimal value function V (see Section 5.5) is C^2, then V must satisfy the differential inequality (5.4).

Problem 5.4 Consider a stock whose price evolution is modeled by the geometric Brownian motion $Y = \{Y_t,\ t \geq 0\}$ that was introduced in Section 5.3. A perpetual American put option is a contract that allows the owner to stop at any time $T \geq 0$ and pays the owner $(S - Y_T)^+$ at the time T chosen, where the *strike price* $S > 0$ is a constant specified in the contract. Mimicking the development in Section 5.3, use the principle of smooth fit to guess an optimal stopping time, and then use Theorem 5.1 to rigorously verify that your guess is correct.

Problem 5.5 Specializing the problem formulation in Section 5.1, suppose that X is standard Brownian motion (zero drift and unit variance) and that $g(x) = \exp(-|x|)$. Show that if $0 < \lambda \leq \frac{1}{2}$ then the stopping time $T := \inf\{t \geq 0 : X_t = 0\}$ is optimal, and that the derivative of its value function $v(\cdot)$ is discontinuous at the origin.

Hint: Because $v'(\cdot)$ has a discontinuity, you cannot use Theorem 5.1 to verify the optimality of T. Rather, you must develop a suitable analog of Theorem 5.1, using Proposition 4.12 (the generalized Itô formula that allows a discontinuous derivative). This same comment applies to the next problem.

Problem 5.6 (*Continuation*) Determine an optimal stopping time when $\lambda > \frac{1}{2}$, and show that the derivative of the associated value function is again discontinuous at the origin.

6

Reflected Brownian Motion

In this chapter we study the stochastic processes (L, U, Z) that are obtained by applying the two-sided reflection mapping to Brownian motion. The role of (L, U, Z) as a storage system model was discussed earlier in Section 2.6. Expected discounted costs will be calculated, and the steady-state distribution of Z will be determined, after certain fundamental properties have been established.

6.1 Strong Markov property

Given parameters μ and $\sigma > 0$, let $(\Omega, \mathcal{F}, \mathbb{F}, P_x)$ be the canonical space described in Section 3.1 and let X be the coordinate process on Ω. Thus X is a (μ, σ) Brownian motion with starting state x on $(\Omega, \mathbb{F}, P_x)$. Now let $b > 0$ be another fixed parameter and let (f, g, h) be the two-sided reflection mapping defined in terms of b in Section 2.4. Restricting attention to starting states $x \in [0, b]$, we define processes $L := f(X)$, $U := g(X)$, and $Z := h(X)$. The definition of the reflection mapping says that

(6.1) L and U are increasing and continuous with $L_0 = U_0 = 0$,

(6.2) $Z_t := X_t + L_t - U_t \in [0, b]$ for all $t \geq 0$, and

(6.3) L and U increase only when $Z = 0$ and $Z = b$, respectively.

In Section 2.4 it was seen that the reflection mapping (f, g, h) has a certain memoryless property. Combining this with the strong Markov property of Brownian motion gives the following important result. By way of setup, let T be an arbitrary stopping time and set

(6.4) $Z_t^* := Z_{T+t}$, $t \geq 0$,

(6.5) $L_t^* := L_{T+t} - L_T$, $t \geq 0$,

and

(6.6) $U_t^* := U_{T+t} - U_T, \qquad t \geq 0$

on $\{T < \infty\}$; these processes will remain undefined on $\{T = \infty\}$. Also, let K be a measurable mapping $(C \times C \times C, \mathcal{C} \times \mathcal{C} \times \mathcal{C}) \to (\mathbb{R}, \mathcal{B})$ such that $E_x(|K(L, U, Z)|) < \infty$ for all $x \in [0, b]$.

Proposition 6.1 *Let* $k(x) := E_x[K(L, U, Z)]$, $0 \leq x \leq b$. *For each* $x \in [0, b]$ *we have*

(6.7) $E_x[K(L^*, U^*, Z^*) \,|\, \mathcal{F}_T] = k(Z_T) \qquad on \, \{T < \infty\}.$

Remark For lack of a better name, (6.7) will be referred to hereafter as the strong Markov property of (L, U, Z). This terminology is not standard.

Proof Fix $x \in [0, b]$. Random variables will be defined only on $\{T < \infty\}$, and identities between random variables will be understood as almost sure relations under P_x. For purposes of this proof, let us define

(6.8) $X_t^* := Z_T + (X_{T+t} - X_T), \qquad t \geq 0.$

From the memoryless property of Proposition 2.7 it follows directly that

(6.9) $L^* = f(X^*), \quad U^* = g(X^*), \quad and \, Z^* = h(X^*).$

That is, the triple (L^*, U^*, Z^*) is obtained by applying the two-sided reflection mapping to X^*. If $y = (y_t, \, t \geq 0)$ is an element of C with $0 \leq y_0 \leq b$, let us set $\Lambda(y) := K(f(y), g(y), h(y))$. Because $L = f(X)$, $U = g(X)$, and $Z = h(X)$, we have

(6.10) $k(x) := E_x[\Lambda(X)].$

Similarly, (6.9) implies that

(6.11) $K(L^*, U^*, Z^*) = \Lambda(X^*)$

and, of course, $Z_T = X_0^*$, so the proposition will be proved if we can establish that

(6.12) $E_x[\Lambda(X^*) \,|\, \mathcal{F}_T] = k(X_0^*).$

Now recall the strong Markov property (3.5) of X. From Theorem 1.7 we have that $X_{T+t} - X_T$ is independent of $\mathcal{F}_T$. Because $Z_T \in \mathcal{F}_T$, it follows that (3.5) continues to hold when X^* is defined by (6.8), although a different definition was used in Section 3.1. Equation (6.12) then follows immediately, because $k(x) := E_x[\Lambda(X)]$. $\square$

In the remainder of this chapter there will be no need to mention explicitly the mapping that carries X into (L, U, Z); we use only the fact that X, L, U, and Z together satisfy (6.1) to (6.3). Thus the letters f, g, and h can and will be reused with new meanings.

6.2 Application of Itô's formula

Fixing $x \in [0, b]$ throughout this section, we set

$$W_t := \frac{1}{\sigma}(X_t - X_0 - \mu t), \qquad t \geq 0.$$

Then W is a standard Brownian motion on the filtered probability space $(\Omega, \mathcal{F}, \mathbb{F}, P_x)$ and $Z := X + L - U$ is an Itô process with Brownian component σW and VF component $\mu t + L - U$. Let $f : [0, b] \to \mathbb{R}$ be C^2. As in Chapters 3 and 4 define the differential operator Γ via

$$\Gamma f := \tfrac{1}{2}\sigma^2 f'' + \mu f'.$$

Proposition 6.2 *$f(Z)$ is an Itô process with differential*

$$df(Z) = \sigma f'(Z)\, dW + [\Gamma f(Z)\, dt + f'(0)\, dL - f'(b)\, dU].$$

Remark The Brownian component of $f(Z)$ has differential $\sigma f'(Z)\, dW$, whereas the quantity in square brackets is the differential of the VF component. Note that the coefficients of dL and dU are constants.

Proof Proceeding exactly as in the proof of Proposition 4.16, we apply Itô's formula to deduce that

$$\begin{aligned}
df(Z) &= f'(Z)\, dZ + \tfrac{1}{2} f''(Z)(dZ)^2 \\
&= f'(Z)(dX + dL - dU) + \tfrac{1}{2} f''(Z)\sigma^2\, dt \\
&= f'(Z)(\sigma\, dW + \mu\, dt + dL - dU) + \tfrac{1}{2}\sigma^2 f''(Z)\, dt \\
&= \sigma f'(Z)\, dW + \Gamma f(Z)\, dt + f'(Z)\, dL - f'(Z)\, dU.
\end{aligned}$$

(6.13)

In its exact integral form, (6.13) says that

$$(6.14) \quad f(Z_t) = f(Z_0) + \sigma \int_0^t f'(Z)\, dW + \int_0^t \Gamma f(Z)\, ds$$
$$+ \int_0^t f'(Z)\, dL - \int_0^t f'(Z)\, dU.$$

But (6.3) gives

$$\int_0^t f'(Z)\, dL = \int_0^t f'(0)\, dL = f'(0)L_t$$

and similarly for $\int f'(Z) \, dU$. Making these substitutions in (6.14) completes the proof. □

Corollary 6.3 *Given* $\lambda > 0$, *set* $u := \lambda f - \Gamma f$ *on* $[0, b]$. *Also, let* $c := f'(0)$ *and* $r := f'(b)$. *Then*

$$(6.15) \qquad f(x) = E_x \left\{ \int_0^\infty e^{-\lambda t} \left[u(Z) \, dt - c \, dL + r \, dU \right] \right\}.$$

Proof Proceeding exactly as in the proof of Proposition 4.17, we use the specialized integration by parts formula of Proposition 4.15 and then Proposition 6.2 to obtain

$$
\begin{aligned}
e^{-\lambda t} f(Z_t) &= f(Z_0) + \int_0^t e^{-\lambda s} \, df(Z) - \lambda \int_0^t e^{-\lambda s} f(Z) \, ds \\
&= f(Z_0) + M_t - \int_0^t e^{-\lambda s} \left[u(Z) \, ds - c \, dL + r \, dU \right]
\end{aligned}
$$

(6.16)

where

$$(6.17) \qquad M_t := \sigma \int_0^t e^{-\lambda s} f'(Z) \, dW.$$

The integrand on the right side of (6.17) is bounded because $0 \le Z \le b$, so $E_x(M_t) = 0$ by Corollary 4.8. Also, $\exp(-\lambda t) f(Z_t) \to 0$ as $t \to \infty$ because $f(Z)$ is bounded, and thus (6.15) is obtained by taking E_x of both sides in (6.16) and letting $t \to \infty$. □

6.3 Expected discounted costs

Hereafter let $\lambda > 0$ be a fixed interest rate. Given a continuous cost rate function u on $[0, b]$ and real constants c and r, we wish to calculate

$$(6.18) \qquad k(x) := E_x \left\{ \int_0^\infty e^{-\lambda t} \left[u(Z) \, dt - c \, dL + r \, dU \right] \right\}.$$

For motivation of this problem, see Section 2.5 and Chapter 7. Corollary 6.3 shows that to compute k one need only solve the ordinary differential equation

$$(6.19) \qquad \lambda k(x) - \Gamma k(x) = u(x), \qquad 0 \le x \le b,$$

with boundary conditions

$$(6.20) \qquad k'(0) = c \quad \text{and} \quad k'(b) = r.$$

Rather than attacking this analytical problem directly, we first use the strong Markov property of (L, U, Z) to obtain a partial solution by probabilistic reasoning. Let $T := T(0) \wedge T(b)$ and define

$$(6.21) \qquad h(x) := E_x \left\{ \int_0^T e^{-\lambda t} u(Z) \, dt \right\}, \qquad 0 \le x \le b.$$

In Section 3.6 we derived a general formula for h in terms of u, observing afterward that

$$(6.22) \qquad \lambda h(x) - \Gamma h(x) = u(x), \qquad 0 \le x \le b,$$

and

$$(6.23) \qquad h(0) = h(b) = 0.$$

Proposition 6.4 *Let $\psi_1(x)$ and $\psi_2(x)$ be defined on $[0, b]$ as in Section 3.3. Then*

$$(6.24) \qquad k(x) = h(x) + \psi_1(x)k(0) + \psi_2(x)k(b), \qquad 0 \le x \le b.$$

Proof We shall apply the strong Markov property (6.7) using the particular functional

$$(6.25) \qquad K(L, U, Z) = \int_0^\infty e^{-\lambda t} [u(Z) \, dt + c \, dL - r \, dU].$$

Taking E_x of both sides in (6.25), it is seen that the definitions of k advanced in Section 6.1 and in this section agree. Comparing (6.18) and (6.21) and using the fact that $L_T = U_T = 0$, we have

$$(6.26) \qquad k(x) = h(x) + E_x \left\{ \int_T^\infty e^{-\lambda t} [u(Z) \, dt + c \, dL - r \, dU] \right\}.$$

Let L^*, U^*, and Z^* be defined as in Section 6.1. Proceeding as in (3.56), but using (6.7) rather than the strong Markov property of X, one finds that

$$
\begin{aligned}
E_x & \left\{ \int_T^\infty e^{-\lambda t} [u(Z) \, dt + c \, dL - r \, dU] \right\} \\
&= E_x \left\{ e^{-\lambda T} \int_0^\infty e^{-\lambda t} [u(Z^*) \, dt + c \, dL^* - r \, dU^*] \right\} \\
&= E_x \left\{ e^{-\lambda T} K(L^*, U^*, Z^*) \right\} \\
&= E_x \left\{ e^{-\lambda T} E_x [K(L^*, U^*, Z^*) | \mathcal{F}_T] \right\} \\
&= E_x \left\{ e^{-\lambda T} k(Z_T) \right\}
\end{aligned}
$$

$$= k(0)E_x(e^{-\lambda T}; Z_T = 0) + k(b)E_x(e^{-\lambda T}; Z_T = b)$$
$$= k(0)\psi_1(x) + k(b)\psi_2(x).$$

Combining this with (6.26) proves the desired identity. □

Explicit formulas for $h(x)$, $\psi_1(x)$, and $\psi_2(x)$ have been derived in Chapter 3, so equation (6.24) reduces our problem to determination of the constants $k(0)$ and $k(b)$. Recall from Section 3.3 that ψ_1 and ψ_2 both satisfy $\lambda\psi - \Gamma\psi = 0$. Thus any function k of the general form (6.24) will satisfy the main equation (6.19), and one simply chooses $k(0)$ and $k(b)$ so as to satisfy the boundary conditions (6.20). An examination of the solutions derived in Chapter 3 will show that (6.24) is equivalent to the general form

$$(6.27) \qquad k(x) = f(x) + Ae^{-\alpha_1(\lambda)x} + Be^{\alpha_2(\lambda)x}$$

where $\alpha_1(\lambda)$ and $\alpha_2(\lambda)$ are the constants defined by (3.22) and (3.23), respectively, A and B are constants to be determined, and

$$(6.28) \qquad f(x) := E_x\left\{\int_0^\infty e^{-\lambda t}u(X_t)\,dt\right\}.$$

In this chapter we have treated u as a function on $[0, b]$; one may use any convenient extension of u for purposes of (6.28). Again it can be verified that any k of the form (6.27) satisfies the main equation (6.19). Therefore one must select A and B so as to meet the boundary conditions (6.20).

6.4 Regenerative structure

Let the starting state $Z(0) = X(0) = x \in [0, b]$ be fixed throughout this section, so we are working with a single filtered probability space $(\Omega, \mathcal{F}, \mathbb{F}, P_x)$. Let

$$(6.29) \qquad T_0 := \inf\{t \geq 0 : Z(t) = 0\}$$

and then for $n = 0, 1, 2, \ldots$ inductively define

$$(6.30) \qquad Z_{n+1}^*(t) := Z(T_n + t), \qquad t \geq 0,$$
$$(6.31) \qquad L_{n+1}^*(t) := L(T_n + t) - L(T_n), \qquad t \geq 0,$$
$$(6.32) \qquad U_{n+1}^*(t) := U(T_n + t) - U(T_n), \qquad t \geq 0,$$

and

$$(6.33) \qquad \begin{aligned} T_{n+1} := \text{ smallest } t > T_n \text{ such that } Z(t) = 0 \text{ and} \\ Z(s) = b \text{ for some } s \in (T_n, t). \end{aligned}$$

In words, T_0 is the first hitting time of level zero, and T_{n+1} is the first time after T_n at which Z returns to level zero after first visiting level b. Then $T_0, T_1, \ldots$ are stopping times, and it follows directly from Proposition 6.1 that, for any $n = 1, 2, \ldots$ and any bounded, measurable $K : C \times C \times C \to \mathbb{R}$,

$$(6.34) \qquad E_x \left[K(L_n^*, U_n^*, Z_n^*) \,|\, \mathcal{F}(T_{n-1}) \right] = E_0 \left[K(L, U, Z) \right].$$

Let $\tau_n := T_n - T_{n-1}$ for $n = 1, 2, \ldots$, and set $\tau := \tau_1$ for ease of notation. It follows from (6.34) that

$$(6.35) \qquad \{\tau_1, \tau_2, \ldots\} \text{ are i.i.d. random variables}$$

and it is left as an exercise (see Problem 6.7) to show that

$$(6.36) \qquad \begin{array}{c} \{\tau_1, \tau_2, \ldots\} \text{ have a nonlattice (or aperiodic or nonarithmetic)} \\ \text{distribution with } E_x(\tau_n) = E_0(\tau) < \infty. \end{array}$$

Conditions (6.34) to (6.36) describe the regenerative structure of our Brownian storage system (L, U, Z). After an initial delay of duration T_0, the *regeneration times* $T_1, T_2, \ldots$ divide the evolution of (L, U, Z) into independent and identically distributed blocks (or regenerative cycles) of duration $\tau_1, \tau_2, \ldots$. Specifically, it follows from (6.34) that

$$(6.37) \qquad \begin{array}{c} \{L_1^*(\tau_1), L_2^*(\tau_2), \ldots\} \quad \text{and} \quad \{U_1^*(\tau_1), U_2^*(\tau_2) \ldots\} \\ \text{are i.i.d. sequences and their distributions do not depend on } x. \end{array}$$

We now define

$$(6.38) \qquad \alpha := \frac{E_0 \left[L(\tau) \right]}{E_0(\tau)}$$

$$(6.39) \qquad \beta := \frac{E_0 \left[U(\tau) \right]}{E_0(\tau)}$$

and

$$(6.40) \qquad \pi(A) := \frac{E_0 \left\{ \int_0^\tau 1_A(Z_t) \, dt \right\}}{E_0(\tau)}$$

for Borel subsets A of $[0, b]$. In words, α and β are the expected increase per time unit in L and U, respectively, over a regenerative cycle. Similarly, $\pi(A)$ is the expected amount of time that Z spends in the set A during a regenerative cycle, normalized to make $\pi(\cdot)$ a probability measure. The following proposition is a standard application of renewal theory, so the proof will only be sketched. See Section 9.2 of Çinlar (1975) for a similar analysis of regenerative processes.

Proposition 6.5   *Let A be an interval subset of [0, b]. Then*

(6.41) $P_x\{Z(t) \in A\} \longrightarrow \pi(A)$ *as* $t \to \infty$,

(6.42) $\frac{1}{t}E_x[L(t)] \longrightarrow \alpha$ *as* $t \to \infty$,

and

(6.43) $\frac{1}{t}E_x[U(t)] \longrightarrow \beta$ *as* $t \to \infty$.

Remark   Thus π, originally defined as an expected occupancy measure during a regenerative cycle, is also the limit distribution of Z, regardless of starting state. In the problems at the end of this chapter, it will be seen that π is also the unique stationary distribution of the Markov process Z and that it may be viewed as a long-run average occupancy distribution. For each of these interpretations of π, there is a corresponding interpretation of α and β, as the problems will show.

Proof   For simplicity, let us assume that $x = 0$, in which case $T_0 = 0$ (the first regenerative cycle begins immediately). To simplify typography, we write $P(\cdot)$ in place of $P_0(\cdot)$. Let $F(t) := P\{\tau \leq t\}$ for $t \geq 0$, noting that F is a nonlattice distribution with

$$a := E(\tau) = \int_0^\infty t\,F(dt) < \infty.$$

First, we have the obvious decomposition

(6.44) $P\{Z(t) \in A\} = P\{Z(t) \in A,\ \tau > t\}$

$$+ \int_0^t P\{\tau \in ds,\ Z(\tau + t - s) \in A\}.$$

From the key condition (6.34) one deduces that

$$\int_0^t P\{\tau \in ds,\ Z(\tau + t - s) \in A\} = \int_0^t P\{\tau \in ds\}P\{Z(t - s) \in A\}$$

(6.45)

$$= \int_0^t P\{Z(t - s) \in A\}\,F(ds).$$

Second, from (6.44), (6.45), and the key renewal theorem (see pages 294–295 of Çinlar, 1975), it follows that

(6.46) $P\{Z(t) \in A\} \longrightarrow \dfrac{1}{a}\displaystyle\int_0^\infty P\{Z(t) \in A,\ T > t\}\,dt.$

Finally, to deduce (6.41) from (6.46), we use Fubini's theorem (see Section A.5) to write

$$\int_0^\infty P\{Z(t) \in A,\ \tau > t\}\, dt = \int_0^\infty E\left[1_{\{Z(t) \in A,\ \tau > t\}}\right] dt$$

$$= E\left[\int_0^\infty 1_{\{Z(t) \in A\}} 1_{\{\tau > t\}}\, dt\right]$$

$$= E\left[\int_0^\tau 1_{\{Z(t) \in A\}}\, dt\right] = a\pi(A).$$

To prove (6.42), first set $Y_n := L_n^*(\tau_n)$ and $S_n := Y_1 + \cdots + Y_n$ for $n = 1, 2, \ldots$, with $S_0 := 0$. Also, let $N(t) = \sup\{n : T_n \leq t\}$ for $t \geq 0$, so $N := \{N(t),\ t \geq 0\}$ is a renewal process with interarrival times $\tau_1, \tau_2, \ldots$. The key observation is that

(6.47) $$S_{N(t)} \leq L(t) \leq S_{N(t)+1} \qquad \text{for } t \geq 0$$

and thus

(6.48) $$\frac{1}{t} E[S_{N(t)}] \leq \frac{1}{t} E[L(t)] \leq \frac{1}{t} E[S_{N(t)+1}].$$

The argument on pages 78–79 of Ross (1983), using Wald's identity and the elementary renewal theorem, shows that both the upper and the lower bounds in (6.48) approach $E(Y_1)/E(\tau_1)$ as $t \to \infty$. Thus $E[L(t)]/t \to E(Y_1)/E(\tau_1)$ as $t \to \infty$, which is precisely (6.42), and (6.43) is established similarly. $\qquad\qquad\qquad\qquad\qquad\qquad\qquad\qquad\qquad\qquad\qquad\qquad$ $\square$

6.5 The steady-state distribution

We now derive a useful relationship, based on Itô's formula, from which one can compute the steady-state quantities $\pi(\cdot)$, α, and β. Let the initial state be $x = 0$, so we are working with the filtered probability space $(\Omega, \mathcal{F}, \mathbb{F}, P_0)$. Proposition 6.2 gives

(6.49) $$f(Z_t) = f(Z_0) + \sigma \int_0^t f'(Z)\, dW + \int_0^t \Gamma f(Z)\, ds + f'(0) L_t - f'(b) U_t$$

for any $f : \mathbb{R} \to \mathbb{R}$ that is C^2. Substituting τ for t in (6.49), we see that $f(Z_\tau) = f(Z_0) = f(0)$. Now take E_0 of both sides. The Itô integral on the right side has expected value zero by Corollary 4.8, because the integrand is bounded and $E(\tau) < \infty$. Thus

(6.50) $$0 = E_0\left\{\int_0^\tau \Gamma f(Z)\, dt\right\} + f'(0) E_0(L_\tau) - f'(b) E_0(U_\tau).$$

Furthermore, from the definition (6.40) of π, it follows that

$$(6.51) \qquad E_0\left\{\int_0^\tau \Gamma f(Z)\,dt\right\} = E_0(\tau)\int_{[0,b]} \Gamma f(z)\,\pi(dz);$$

this relationship holds by definition if Γf is the indicator of a set, then by linearity it holds whenever Γf is a simple function (linear combination of indicators), and then by monotone convergence it holds in general. Finally, $E_0(L_\tau) = \alpha E_0(\tau)$ and $E_0(U_\tau) = \beta E_0(\tau)$ by definition. Substituting these identities and (6.51) into (6.50), then dividing by $E_0(\tau)$, one arrives at the key relationship

$$(6.52) \qquad 0 = \int_{[0,b]} \Gamma f(z)\,\pi(dz) + \alpha f'(0) - \beta f'(b).$$

Proposition 6.6 *If $\mu = 0$, then $\alpha = \beta = \sigma^2/2b$ and π is the uniform distribution on $[0, b]$. Otherwise, setting $\theta := 2\mu/\sigma^2$,*

$$(6.53) \qquad \alpha = \frac{\mu}{e^{\theta b} - 1}, \qquad \beta = \frac{\mu}{1 - e^{-\theta b}},$$

and π is the truncated exponential distribution

$$(6.54) \qquad \pi(dz) = p(z)\,dz \qquad where\ p(z) = \frac{\theta e^{\theta z}}{e^{\theta b} - 1}.$$

Proof First, suppose $\mu = 0$. Substitute into (6.52) the linear function $f(z) = z$. Then $\Gamma f = 0$, $f'(0) = f'(b) = 1$, and (6.52) gives $\alpha = \beta$. Next, take $f(z) = z^2$ so that $\Gamma f(z) = \sigma^2$, $f'(0) = 0$, and $f'(b) = 2b$. Then (6.52) yields $\alpha = \beta = \sigma^2/2b$. Finally, consider the exponential function $f(z) = \exp(\lambda z)$. Substituting this into (6.52) and using the known values of α and β, we arrive at

$$\int_{[0,b]} e^{\lambda y}\,\pi(dy) = \frac{1}{b\lambda}(e^{b\lambda} - 1), \qquad \lambda \in \mathbb{R},$$

which is the transform of the uniform distribution as desired. If μ is nonzero, substitution of the test function $f(z) = z$ in (6.52) gives

$$(6.55) \qquad 0 = \mu + \alpha - \beta.$$

Now consider again the exponential test function $f(z) = \exp(\lambda z)$ so that

$$(6.56) \qquad \Gamma f(z) = \left(\tfrac{1}{2}\sigma^2\lambda^2 + \mu\lambda\right)e^{\lambda z}$$

$$(6.57) \qquad f'(0) = \lambda \quad \text{and} \quad f'(b) = \lambda e^{\lambda b}.$$

By taking $\lambda = -2\mu/\sigma^2 = -\theta$, we have $\Gamma f = 0$, and hence (6.52) yields

(6.58) $$0 = -\theta\alpha + \theta\beta e^{-\theta b}.$$

Solving (6.55) and (6.58) simultaneously gives (6.53). Now let us return to general λ. Using (6.56), (6.57), and (6.53) in (6.52), we arrive at

$$\int_{[0,b]} e^{\lambda z}\, \pi(dz) = \left(\frac{\theta}{\theta + \lambda}\right)\left[\frac{e^{(\theta+\lambda)b} - 1}{e^{\theta b} - 1}\right]$$

which is the transform of the truncated exponential distribution (6.54), as desired.   □

An important quantity in applications is the mean of the steady-state distribution (6.54). Readers may verify that

(6.59) $$\gamma := \int_0^b zp(z)\, dz = \frac{b}{1 - e^{-\theta b}} - \frac{1}{\theta}.$$

To express the system performance measures α, β, γ in more compact form, let

(6.60) $$\psi(\xi) = \frac{\xi}{e^{\xi} - 1}$$

and

(6.61) $$\phi(\xi) := \frac{e^{-\xi} - 1 + \xi}{\xi(1 - e^{-\xi})}.$$

with $\psi(0) = 1$ and $\phi(0) = \frac{1}{2}$. (These are the values that make ψ and ϕ continuous at the origin.) It has been shown in this section that

(6.62) $$\alpha = \left(\frac{\sigma^2}{2b}\right)\psi\left(\frac{2\mu b}{\sigma^2}\right),$$

(6.63) $$\beta = \alpha + \mu,$$

and

(6.64) $$\gamma = b\phi\left(\frac{2\mu b}{\sigma^2}\right).$$

Incidentally, it follows from (6.41) that

(6.65) $$E_x(Z_t) \longrightarrow \gamma \qquad \text{as } t \to \infty.$$

6.6 The case of a single barrier

Assuming that $X_0 = x \geq 0$, let us now consider the processes (L, Z) obtained by applying to X the one-sided reflection mapping of Section 2.2. (Recall that the distribution of Z_t was calculated explicitly for general values of t and x in Section 3.7.) Each of the results developed in Sections 5.1 to 5.5 has a precise analog in the case of a single barrier, and the most important of these will be recorded here with the proofs left as exercises. Recall from Section 2.2 that

(6.66) L is increasing and continuous with $L_0 = 0$,

(6.67) $Z_t := X_t + L_t \geq 0$ for all $t \geq 0$, and

(6.68) L increases only when $Z = 0$.

Thus Z is an Itô process with Brownian component σW and VF component $\mu t + L$. Using (6.68) and Itô's formula, one finds that

(6.69) $$f(Z_t) = f(Z_0) + \sigma \int_0^t f'(Z)\, dW + \int_0^t \Gamma f(Z)\, dt + f'(0)L_t$$

for any $f : \mathbb{R} \to \mathbb{R}$ that is C^2. If f also has bounded derivative, then it follows from (6.69) that, for any $\lambda > 0$,

(6.70) $$f(x) = E_x\left\{ \int_0^\infty e^{-\lambda t} [u(Z)\, dt - c\, dL] \right\}$$

where

(6.71) $$u(x) := \lambda f(x) - \Gamma f(x) \quad \text{and} \quad c := f'(0).$$

Turning this calculation around, suppose there is given a constant c and a bounded, continuous cost rate function $u : [0, \infty) \to \mathbb{R}$. If we wish to calculate the expected discounted cost

(6.72) $$k(x) := E_x\left\{ \int_0^\infty e^{-\lambda t} [u(Z)\, dt - c\, dL] \right\}, \qquad x \geq 0,$$

it suffices to solve the differential equation $\lambda k(x) - \Gamma k(x) = u(x)$, $x \geq 0$, subject to the requirement that $k'(0) = c$ and $k'(\cdot)$ is bounded on $[0, \infty)$. Imitating the arguments in Section 6.1 and Section 6.3, it can be shown that

(6.73) $$k(x) = g(x) + k(0)e^{-\alpha_1(\lambda)x}, \qquad x \geq 0,$$

where

(6.74) $$g(x) := E_x\left\{\int_0^T e^{-\lambda t}u(X_t)\,dt\right\}, \qquad x \geq 0,$$

and $T := \inf\{t \geq 0 : X_t = 0\}$. A general formula for g was derived in Section 3.5, and it follows from the results of Sections 3.3 to 3.5 that any function k of the form (6.73) satisfies $\lambda k - \Gamma k = u$ on $[0, \infty)$. Moreover, boundedness of u implies boundedness of g', and the boundary condition $k'(0) = c$ can be satisfied by taking

(6.75) $$k(0) = \frac{g'(0) - c}{\alpha_1(\lambda)};$$

thus formulas (6.73) to (6.75) provide a complete solution of the problem at hand. For simplicity, this treatment of expected discounted costs has been restricted to bounded cost rate functions u, but the solution (6.73) to (6.75) remains valid so long as the expectation in (6.74) makes sense, as one can show with a truncation argument.

Asymptotic analysis of Z is much easier with one barrier than with two, because we have previously calculated the distribution of Z_t for finite t. Specifically, letting $t \to \infty$ in formula (3.63), one finds that (for any $z \geq 0$)

(6.76) $$P_x\{Z_t \leq z\} \longrightarrow 1 - \exp\left(\frac{2\mu z}{\sigma^2}\right) \qquad \text{if } \mu < 0$$

whereas $P_x\{Z_t \leq z\} \to 0$ if $\mu \geq 0$ as one would expect. Note that the exponential limit distribution in (6.76) is what one gets by simply letting $b \to \infty$ in the steady-state distribution calculated earlier in Section 6.5. From formula (3.63) it also follows that

(6.77) $$E_x(Z_t) \longrightarrow \frac{\sigma^2}{2|\mu|} \qquad \text{as } t \to \infty \quad \text{if } \mu < 0,$$

which is what one would expect from (6.76). Using the fact that $E_x(X_t) = x + \mu t$, we can take E_x of both sides in (6.67) to obtain

(6.78) $$E_x(Z_t) = x + \mu t + E_x(L_t).$$

Now divide (6.78) by t, let $t \to \infty$, and use (6.77) to conclude that

(6.79) $$\frac{1}{t}E_x(L_t) \longrightarrow |\mu| \qquad \text{as } t \to \infty \quad \text{if } \mu < 0.$$

Readers should note that the constant α computed earlier in Section 6.5 approaches $|\mu|$ as $b \to \infty$ if $\mu < 0$, which is consistent with (6.79).

6.7 Problems and complements

Problem 6.1   In the setting of Section 6.2, let $T := \inf\{t \geq 0 : Z_t = b\}$ and note that $U(T) = 0$. Let $f : \mathbb{R} \to \mathbb{R}$ be C^2. Use Proposition 6.2 to prove that

$$E_x\left[f\left(Z(t \wedge T)\right)\right] = f(x) + E_x\left[\int_0^{t \wedge T} \Gamma f(Z)\, ds\right] + f'(0)E_x\left[L(t \wedge T)\right].$$

Specializing to $f(x) := \exp(\lambda x)$, note that $f'(0) = \lambda$, and $\Gamma f(x) = q(\lambda)f(x)$, where $q(\lambda) = \frac{1}{2}\sigma^2\lambda^2 + \mu\lambda$. Choosing $\lambda > 0$ large enough to ensure $q(\lambda) > 0$, show that

(6.80) $$E_x\left[f\left(Z(t \wedge T)\right)\right] \geq f(x) + q(\lambda)E_x(t \wedge T) + \lambda E_x\left[L(t \wedge T)\right].$$

But $f(Z(t \wedge T)) \leq f(b)$, and $L(\cdot) \geq 0$, so $E_x(t \wedge T) \leq [f(b) - f(x)]/q(\lambda)$ by (6.80). Now let $t \uparrow \infty$ and use the monotone convergence theorem to conclude that $E_x(T) < \infty$, $0 \leq x \leq b$.

Problem 6.2 (*Continuation*)   Let f again be general. Use Proposition 6.2 and Corollary 4.8 to show that

$$f(b) = f(x) + E_x\left[\int_0^T \Gamma f(Z)\, dt\right] + f'(0)E_x\left[L(T)\right], \qquad 0 \leq x \leq b.$$

Problem 6.3 (*Continuation*)   Let $\phi(x) := E_x[\exp(-\lambda T)]$ where $\lambda > 0$. Show that

$$f(b)\phi(x) = f(x) + E_x\left\{\int_0^T e^{-\lambda t}\left[(\Gamma f - \lambda f)(Z)\, dt + f'(0)\, dL\right]\right\}.$$

Problem 6.4 (*Continuation*)   Let $\alpha_1(\lambda)$ and $\alpha_2(\lambda)$ be defined as in Section 3.3 and set

$$g(x) := \alpha_2(\lambda)e^{-\alpha_1(\lambda)x} + \alpha_1(\lambda)e^{\alpha_2(\lambda)x}, \qquad 0 \leq x \leq b.$$

From the results of Section 3.3 it follows that $\lambda g - \Gamma g = 0$ on $[0, b]$ and clearly $g'(0) = 0$. Conclude that $\phi(x) = g(x)/g(b)$, $0 \leq x \leq b$.

Problem 6.5   Again consider the setup of Section 6.2. Fix $\kappa > 0$, let $f : [0, b] \to \mathbb{R}$ be C^2, and let

$$V_t := e^{-\kappa L_t} f(Z_t), \qquad t \geq 0.$$

Use the integration by parts formula (4.55) to calculate the differential of the Itô process V.

Problem 6.6 (*Continuation*) Now let $f(x) := E_x\{\int_0^T e^{-\kappa L_t} u(Z_t)\,dt\}$, $0 \le x \le b$, where u is continuous on $[0, b]$ and T is the first hitting time of b as in Problems 6.1 to 6.4. Show that to compute f it suffices to solve the differential equation

$$\Gamma f + u = 0$$

with boundary conditions

$$f'(0) - \kappa f(0) = 0 \quad \text{and} \quad f(b) = 0.$$

Problem 6.7 In the setting of Section 6.4, let $T(y) := \inf\{t \ge 0 : Z_t = y\}$. In Problem 6.1 it was shown that $E_x[T(b)] < \infty$, $0 \le x \le b$, and essentially the same argument gives $E_x[T(0)] < \infty$. Use Proposition 6.1 to show that $E_0(\tau) = E_0[T(b)] + E_b[T(0)] < \infty$.

Problem 6.8 In the setting of Section 6.4 (where the starting state x is viewed as a fixed constant), it can be shown that

$$\frac{1}{t}\int_0^t 1_A(Z)\,ds \longrightarrow \pi(A) \qquad \text{almost surely as } t \to \infty,$$

$$\frac{1}{t}L(t) \longrightarrow \alpha \qquad \text{almost surely as } t \to \infty,$$

and

$$\frac{1}{t}U(t) \longrightarrow \beta \qquad \text{almost surely as } t \to \infty.$$

For this one uses the regenerative structure of (L, U, Z), the standard form of the strong law of large numbers and the strong law for renewal processes. A very similar argument can be found on page 78 of Ross (1983).

Problem 6.9 A probability measure π on $[0, b]$ is said to be a stationary distribution for Z if

(6.81) $$\int_{[0,b]} \pi(dx)E_x[f(Z_t)] = \int_{[0,b]} \pi(dz)f(z)$$

for all $t \ge 0$ and all bounded, measurable f. Condition (6.81) says that if the initial state of Z is randomized with distribution π, then Z_t has distribution π at each future time t. Use (6.7) to show that if π is a stationary distribution for Z, then

$$(6.82) \qquad \int_{[0,b]} \pi(dx)E_x(L_t) = \alpha t \quad \text{and} \quad \int_{[0,b]} \pi(dx)E_x(U_t) = \beta t$$

for all $t \geq 0$, where α and β are constants yet to be determined.

Problem 6.10 (*Continuation*) Let f be bounded and measurable on $[0, b]$. One immediate consequence of (6.7) is that

$$E_x[f(Z_{t+s})] = \int_{[0,b]} P_x\{Z_t \in dy\}E_y[f(Z_s)].$$

Letting $t \to \infty$, use the bounded convergence theorem to conclude that the limit distribution calculated in Section 6.5 is also a stationary distribution for Z. Now to prove uniqueness, let π be any stationary distribution. From Proposition 6.2 it follows that

$$E_x[f(Z_t)] = f(x) + E_x\left[\int_0^t \Gamma f(Z)\,ds + f'(0)L_t - f'(b)U_t\right]$$

for any $t \geq 0$, $x \in [0, b]$, and C^2 function f. Integrate both sides of this equation with respect to $\pi(dx)$, then use (6.81) and (6.82) to show that π, α, and β jointly satisfy (6.52) for all C^2 test functions f. Thus π, α, and β are the same quantities computed in Section 6.5.

Problem 6.11 Consider the setup of Section 6.6, where $Z := X + L$ is a reflected (μ, σ) Brownian motion with a single barrier at zero. Let $f(t, x)$ be twice continuously differentiable on $\mathbb{R}^2$, and define

$$g(t, x) := \frac{\partial}{\partial x} f(t, x)$$

and

$$h(t, x) := \left(\frac{\partial}{\partial t} + \frac{1}{2}\sigma^2 \frac{\partial^2}{\partial x^2} + \mu \frac{\partial}{\partial x}\right) f(t, x).$$

Use the multi-dimensional Itô formula (4.38) to show that

$$f(t, Z_t) = f'(0, Z_0) + \sigma \int_0^t g(s, Z_s)\,dW + \int_0^t h(s, Z_s)\,ds + \int_0^t g(s, 0)\,dL.$$

Now suppose that $g(s, y)$ is bounded on $[0, t] \times [0, \infty)$, that $h(s, y) = 0$ on $[0, t] \times [0, \infty)$, and that $g(s, 0) = 0$ on $[0, t]$. Use Corollary 4.8 to show that $f(0, x) = E_x[f(t, Z_t)]$ for $x \geq 0$.

Problem 6.12 (*Continuation*)   Fix $y \geq 0$ and let $Q(t, x, y)$ be defined by formula (3.63), recalling that

$$\frac{\partial}{\partial t} Q(t, x, y) = \left(\frac{1}{2} \sigma^2 \frac{\partial^2}{\partial x^2} + \mu \frac{\partial}{\partial x} \right) Q(t, x, y),$$

$$\frac{\partial}{\partial x} Q(t, 0, y) = 0, \quad \text{and} \quad Q(0, x, y) = 1_{(x>y)}.$$

It is also easy to check that $(\partial/\partial x)Q(t, x, y)$ is bounded as a function of t and x. Use the result of Problem 6.11 to prove that $Q(t, x, y) = P_x\{Z_t > y\}$, thus verifying the interpretation of Q given in Section 3.7. This requires a sequence of steps exactly like those outlined in Problems 4.10 to 4.13.

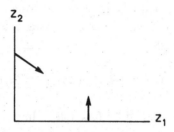

Figure 6.1 Directions of reflection for Z.

Problem 6.13   It is the purpose of this problem to give some idea of the role played by stochastic calculus in the analysis of multi-dimensional Brownian storage models. Consider the tandem buffer system discussed earlier in Problem 2.3. Suppose that the netflow process $X = (X_1, X_2)$ is modeled as a two-dimensional standard Brownian motion. (This means that X_1 and X_2 are independent, each with zero drift and unit variance. Similar results are obtained with arbitrary drift vector and covariance matrix.) Let S (for *state space*) denote the positive quadrant of $\mathbb{R}^2$. We extend our previous notational system to denote by P_x the probability measure on the path space of X corresponding to starting state $x = (x_1, x_2) \in S$. Applying to X the multi-dimensional reflection mapping of Problem 2.3, one obtains processes $L = (L_1, L_2)$ and $Z = (Z_1, Z_2)$ satisfying

(6.83) L_1 and L_2 are increasing and continuous with $L_1(0) = L_2(0) = 0$,

(6.84) $Z_1(t) := X_1(t) + L_1(t) \geq 0$ for all $t \geq 0$,

(6.85) $Z_2(t) := X_2(t) - L_1(t) + L_2(t) \geq 0$ for all $t \geq 0$, and

(6.86) L_1 and L_2 increase only when $Z_1 = 0$ and $Z_2 = 0$, respectively.

Recall that the path-to-path mapping that carries X into (L, Z) is naturally described by the *directions of reflection* shown in Figure 6.1. From (6.83) to (6.85) we see that Z is a two-dimensional Itô process. Now let $f : \mathbb{R}^2 \to \mathbb{R}$ be twice continuously differentiable and define the differential operators (here subscripts denote partial derivatives as in Section 4.7)

$$(6.87) \qquad \Delta f := f_{11} + f_{22},$$

$$(6.88) \qquad D_1 f := f_1 - f_2,$$

and

$$(6.89) \qquad D_2 f := f_2.$$

Thus D_1 and D_2 are directional derivatives for the directions of reflection associated with the boundary surfaces $Z_1 = 0$ and $Z_2 = 0$, respectively. Apply the multi-dimensional Itô formula to show that

$$(6.90) \qquad \begin{aligned} df(Z) &= \sum_{i=1}^{2} f_i(Z) \, dZ_i + \frac{1}{2} \sum_{i=1}^{2} \sum_{j=1}^{2} f_{ij}(Z) \, dZ_i \, dZ_j \\ &= \sum_{i=1}^{2} f_i(Z) \, dX_i + \frac{1}{2} \Delta f(Z) \, dt + \sum_{i=1}^{2} D_i f(Z) \, dL_i. \end{aligned}$$

This provides a precise analog for Proposition 6.2, and proceeding as in Section 6.2, we can use (6.90) and Proposition 4.15 to obtain

$$(6.91) \quad e^{-\lambda T} f(Z(T)) = f(Z(0)) + \sum_{i=1}^{2} \int_0^T e^{-\lambda t} f_i(Z) \, dX_i$$

$$+ \int_0^T e^{-\lambda t} \left(\frac{1}{2} \Delta f - \lambda f \right)(Z) \, dt + \sum_{i=1}^{2} \int_0^T e^{-\lambda t} D_i f(Z) \, dL_i$$

for any constant $\lambda > 0$ and stopping time $T < \infty$. Let T be a fixed time, suppose f and its first-order partials are bounded on S, and take E_x of both sides in (6.91). The stochastic integrals have expected value zero by Corollary 4.8, and upon letting $T \to \infty$ we arrive at

$$(6.92) \qquad f(x) = E_x \left\{ \int_0^\infty e^{-\lambda t} \left[\left(\lambda f - \frac{1}{2} \Delta f \right)(Z) \, dt - \sum_{i=1}^{2} D_i f(Z) \, dL_i \right] \right\}.$$

Given constants c_1, c_2, and a well-behaved cost rate function $u : S \to \mathbb{R}$, suppose we wish to calculate

$$(6.93) \qquad k(x) := E_x \left\{ \int_0^\infty e^{-\lambda t} [u(Z) \, dt - c_1 \, dL_1 - c_2 \, dL_2] \right\}.$$

Using (6.92), show that it suffices to find a sufficiently regular function k satisfying the partial differential equation

(6.94) $\frac{1}{2}\Delta k(x) - \lambda k(x) + u(x) = 0$, $x \in S$,

subject to the boundary conditions

(6.95) $D_1 k(0, x_2) = c_1$, $x_2 \geq 0$

and

(6.96) $D_2 k(x_1, 0) = c_2$, $x_1 \geq 0$.

Justification of the boundary conditions depends critically on the sample path property (6.86). For more on the theory of multi-dimensional reflected Brownian motion, see Harrison and Reiman (1981).

7

Optimal Control of Brownian Motion

In a stochastic control problem one observes and then seeks to favorably in-
fluence the behavior of some stochastic system. Such problems involve *dynamic* optimization, meaning that observations and actions are spread out
in time. In this chapter several simple but fundamental stochastic control
problems will be solved directly from first principles, with heavy reliance
on the ubiquitous Itô formula. For each of the closely related problems to
be considered, the optimal policy involves the imposition of control barriers; these may be either jump barriers or reflecting barriers, depending on
the cost structure assumed.

Our first problem can be informally described as follows. Consider a
controller who continuously monitors the contents of a storage system,
such as an inventory or a bank account. In the absence of control, the contents process $Z = \{Z_t, t \geq 0\}$ fluctuates as a (μ, σ) Brownian motion. The
controller can at any time increase or decrease the contents of the system
by any amount desired, but is obliged to keep $Z_t \geq 0$, and there are three
types of costs to be considered. First, to increase the contents from x to
$x + \delta$, the controller must pay a fixed charge K plus a proportional charge
$k\delta$. Similarly, it costs $L + \ell\delta$ to decrease the contents from x to $x - \delta$. Finally, inventory holding costs are continuously incurred at rate κZ_t. Thus
we have a problem with *linear holding costs* and *fixed-plus-proportional
costs of control*. Section 7.1 contains a precise mathematical statement of
this problem, assuming that future costs are continuously discounted at interest rate $\lambda > 0$. That discounted cost problem is solved in Sections 7.2
and 7.3, and then Sections 7.4 to 7.6 consider its undiscounted analog, in
which the controller's objective is to minimize long-run average cost per
time unit.

For problems of the kind just described, control takes the form of positive and negative jumps that are enforced at isolated points in time. These
are called *impulse control problems*. We shall also consider the problem
variant in which $K = L = 0$, the precise formulation of which involves

some subtlety. This problem will be analyzed with a discounted cost criterion in Section 7.7, and then with a long-run average cost criterion in Section 7.8. In each case the optimal policy takes the form of two reflecting barriers: a lower barrier at zero and an upper barrier at an appropriate level $b > 0$. Under such a policy, control is exerted at uncountably many time points, but the total amount of upward or downward displacement effected in any finite time interval is finite. Following the terminology of Harrison and Taksar (1983), we shall describe the alternative formulation with only proportional control costs as an *instantaneous control problem*. To conclude the chapter, Section 7.9 discusses applications to cash management.

7.1 Impulse control with discounting

Given parameters μ and $\sigma > 0$, let $(\Omega, \mathcal{F}, \mathbb{F}, P_x)$ be the canonical space described in Section 3.1, and let X be the coordinate process on Ω. Thus X is a (μ, σ) Brownian motion with starting state x on $(\Omega, \mathcal{F}, \mathbb{F}, P_x)$, and attention is restricted to starting states $x \geq 0$. Other data for our problem are fixed control costs $K, L > 0$, proportional cost rates $k, \ell > 0$, a holding cost rate $\kappa > 0$, and an interest rate $\lambda > 0$ for discounting. We assume λ is small enough to ensure that

(7.1) $\kappa/\lambda > \ell.$

The left side of (7.1) is the total discounted cost of holding one unit of inventory forever, and the right side is the variable cost of removing one unit of inventory. Thus (7.1) eliminates from consideration an uninteresting degenerate case: if it did not hold, then an optimal policy would never exercise downward control, no matter how large the system contents might become.

A *policy* consists of a sequence of stopping times $\{T_0, T_1, \ldots\}$ and a sequence of random variables $\{\xi_0, \xi_1, \ldots\}$ such that

$$P_x(0 = T_0 < T_1 < \cdots \rightarrow \infty) = 1, \qquad \text{for all } x \in \mathbb{R},$$

and

$$\xi_n \in \mathcal{F}_{T_n}, \qquad \text{for all } n = 0, 1, \ldots .$$

Interpret T_n as the nth time at which the controller enforces a jump in the state of the system, with ξ_n the size of the jump (either positive or negative) enforced. The convention $T_0 = 0$ will prove to be convenient, but then we must allow $\xi_0 = 0$. We associate with a policy $\{(T_n, \xi_n)\}$ the processes

$$
\begin{aligned}
N(t) &= \sup\{n \geq 0 : T_n \leq t\}, & t &\geq 0, \\
Y_t &= \xi_1 + \cdots + \xi_{N(t)}, & t &\geq 0, \\
Z_t &= X_t + Y_t, & t &\geq 0.
\end{aligned}
$$

Note that N, Y, and Z are all adapted and right-continuous with left limits. The policy $\{(T_n, \xi_n)\}$ is said to be *feasible* if

$$(7.2) \qquad P_x(Z_t \geq 0 \text{ for all } t \geq 0) = 1, \qquad \text{for all } x \geq 0,$$

$$(7.3) \qquad \sup_{t \geq 1} \frac{1}{t} E_x \left[\sum_{n=0}^{N(t)} (1 + |\xi_n|) \right] < \infty \qquad \text{for all } x \geq 0,$$

and there exists a finite constant $A > 0$ such that

$$(7.4) \qquad P_x \left(\sup_{t \geq 0} Z_t < A \right) = 1 \qquad \text{for all } x \geq 0.$$

By restricting attention from the outset to policies that satisfy (7.3) and (7.4) we avoid uninteresting technical complexities. Furthermore, it will be obvious from the development to follow that any policy which violates (7.3) or (7.4) can be dominated, in the sense of expected discounted cost, by another policy which satisfies those restrictions. Next, setting

$$(7.5) \qquad \phi(\xi) = \begin{cases} K + k\xi & \text{if } \xi > 0 \\ 0 & \text{if } \xi = 0 \\ L - \ell\xi & \text{if } \xi < 0, \end{cases}$$

we define the *cost function* for a feasible policy $\{(T_n, \xi_n)\}$ by

$$
G(x) = E_x \left[\int_0^\infty e^{-\lambda t} \kappa Z_t \, dt + \sum_{n=0}^\infty e^{-\lambda T_n} \phi(\xi_n) \right]
$$

for all $x \geq 0$. From (7.3) and (7.4) it follows that $G(x)$ is both well defined and finite for all $x \geq 0$. We say that this policy is *optimal* if $G(x)$ is minimal, among the cost functions associated with feasible policies, for each $x \geq 0$.

Proposition 7.1 *Suppose that* $f : [0, \infty) \rightarrow \mathbb{R}$ *is* C^1, *has a bounded derivative, and is piecewise* C^2 *(see Guide to Notation and Terminology for the precise meaning of that term). Then for any fixed time* $T > 0$, *any* *x ≥ 0, and any feasible policy we have*

$$(7.6) \quad E_x\left[e^{-\lambda T} f(Z_T)\right] = E_x\left[f(Z_0)\right] + E_x\left[\int_0^T e^{-\lambda t}(\Gamma f - \lambda f)(Z_t)\,dt\right.$$

$$\left. + E_x\left[\sum_{n=1}^{N(T)} \theta_n e^{-\lambda T_n}\right],\right.$$

where

$$(7.7) \qquad \theta_n := f(Z(T_n)) - f(Z(T_n-)) \qquad for\ n = 1,2,\ldots$$

and $\Gamma f := \frac{1}{2}\sigma^2 f'' + \mu f'$.

First Remark We may define $f''(y)$ arbitrarily at those points y where the second derivative does not exist; see comments following Proposition 4.12.

Second Remark This proposition is also valid for $\lambda = 0$. That fact will be used later in Sections 7.4, 7.5, and 7.8.

Proof As in Chapter 4, we can represent X as $X_t = X_0 + \mu t + \sigma W_t$, where W is a standard Brownian motion. Thus the contents process Z for an arbitrary feasible policy satisfies $dZ = \mu\,dt + \sigma\,dW + dY$. Proceeding as in the proof of Proposition 4.17, we apply Proposition 4.13 (the generalization of Itô's formula that allows jumps in the process Z) and Proposition 4.15 (the integration by parts formula with one factor a negative exponential) to obtain the following:

$$(7.8) \quad e^{-\lambda T} f(Z_T) = f(Z_0) + M_T + \int_0^T e^{-\lambda t}(\Gamma f - \lambda f)(Z_t)\,dt + \sum_{n=1}^{N(T)} \theta_n e^{-\lambda T_n},$$

where θ_n is defined by (7.7) and

$$M_T := \sigma \int_0^T e^{-\lambda t} f'(Z_t)\,dW_t.$$

Taking E_x of both sides of (7.8), we have $E_x(M_T) = 0$ by Corollary 4.8 (the zero expectation property of the stochastic integral), because $f'(\cdot)$ is bounded by hypothesis, and that establishes the desired conclusion. □

Proposition 7.2 *Suppose that* $f : [0,\infty) \to \mathbb{R}$ *satisfies all the hypotheses* *of Proposition 7.1 plus*

$$(7.9) \qquad \lambda f(x) - \Gamma f(x) \le \kappa x \qquad\qquad \text{for almost all } x \ge 0,$$

$$(7.10) \qquad f(x) - f(y) \le K + k(y - x) \qquad \text{for } 0 \le x < y,$$

$$(7.11) \qquad f(x) - f(y) \le L - \ell(y - x) \qquad \text{for } 0 \le y < x.$$

Then $f(x) \le G(x)$ for all $x \ge 0$, where $G(\cdot)$ is the cost function for any feasible policy.

Proof Let us consider an arbitrary feasible policy, consisting of jump times $\{T_n\}$ and jump sizes $\{\xi_n\}$; other processes and functions associated with the policy, including its cost function $G(\cdot)$, are defined as in the text preceding Proposition 7.1. Using the definition (7.5) of $\phi(\cdot)$, we see that (7.10) and (7.11) together imply $f(x) - f(y) \le \phi(y - x)$, which means that

$$(7.12) \qquad -\theta_n \le \phi((Z(T_n) - Z(T_n-))) = \phi(\xi_n) \qquad \text{for } n = 1, 2, \dots$$

where θ_n is defined by (7.7). Substituting (7.9) and (7.12) into (7.6) and rearranging terms, we have

$$(7.13) \quad E_x[f(Z_0)] \le E_x \left[\sum_{n=1}^{N(T)} e^{-\lambda T_n} \phi(\xi_n) \right] + E_x \left[\int_0^T e^{-\lambda t} \kappa Z_t \, dt \right]$$
$$+ E_x \left[e^{-\lambda T} f(Z_T) \right].$$

From (7.4) it follows that the last term on the right side of (7.13) vanishes as $T \to \infty$, so we have

$$(7.14) \qquad E_x[f(Z_0)] \le E_x \left[\sum_{n=1}^{\infty} e^{-\lambda T_n} \phi(\xi_n) \right] + E_x \left[\int_0^{\infty} e^{-\lambda t} \kappa Z_t \, dt \right].$$

Finally, because $Z_0 = X_0 + \xi_0$, another application of (7.10) and (7.11) gives

$$(7.15) \qquad\qquad f(X_0) \le f(Z_0) + \phi(\xi_0).$$

Of course, $E_x[f(X_0)] = f(x)$, so combining (7.14) and (7.15) we have

$$f(x) \le E_x \left[\sum_{n=0}^{\infty} e^{-\lambda T_n} \phi(\xi_n) \right] + E_x \left[\int_0^{\infty} e^{-\lambda t} \kappa Z_t \, dt \right] := G(x). \qquad \square$$

7.2 Control band policies

For the impulse control problem with discounting, it is plausible that there exists an optimal policy of the form pictured in Figure 7.1. This *control band policy* is specified by three parameters (q, Q, S) satisfying $0 < q <$

$Q < S$. Whenever the contents process Z hits zero, the controller enforces an upward jump to level q, incurring a control cost of $K + kq$. On the other hand, whenever Z reaches the upper limit S, the controller enforces a downward jump to level Q, incurring a control cost of $L + \ell(S - Q)$. If the initial contents X_0 exceeds S, the controller immediately enforces a downward jump to Q. Thus we have

$$\xi_0 = \begin{cases} q & \text{if } X_0 = 0 \\ 0 & \text{if } 0 < X_0 < S \\ Q - X_0 & \text{if } X_0 \geq S. \end{cases}$$

Assuming it is clear from the verbal description how $T_1, \xi_1, T_2, \xi_2, \ldots$ are constructed recursively for a control band policy, their formal definitions are omitted.

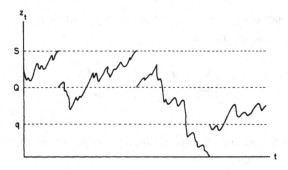

Figure 7.1 A control band policy.

Let us now consider the cost function $G(x)$ for a control band policy with parameters (q, Q, S), assuming initially that $0 \leq x \leq S$. Defining T as the first $t \geq 0$ at which either $X_t = 0$ or $X_t = S$, we decompose $G(x)$ into the expected present value of costs incurred strictly before time T (which consist entirely of inventory holding costs), plus the expected present value of costs incurred at time T and afterward. Using the strong Markov property of X as in Problem 3.13, that decomposition can be expressed as follows:

$$(7.16) \qquad G(x) = h(x) + \psi_1(x)G(0) + \psi_2(x)G(S), \qquad 0 \leq x \leq S,$$

where $h(x)$ is defined as in Section 3.6 but with S in place of b and with the specific cost rate function $u(x) = \kappa x$, and $\psi_1(x)$ and $\psi_2(x)$ are defined as in Section 3.3, again with S in place of b. Now $h(x)$ satisfies the differential equations $\lambda h(x) - \Gamma h(x) = \kappa x$ (see Section 3.6), while $\psi_1(x)$ and $\psi_2(x)$ each satisfy $\lambda\psi(x) - \Gamma\psi(x) = 0$ (see Section 3.3), so (7.16) implies that

(7.17) $\qquad\qquad \lambda G(x) - \Gamma G(x) = \kappa x, \qquad 0 \le x \le S,$

to which we add the following obvious boundary conditions:

(7.18) $\qquad\qquad G(0) = K + kq + G(q),$

(7.19) $\qquad\qquad G(S) = L + \ell(S - Q) + G(Q).$

Actually, the more general version of (7.19) is

(7.20) $\qquad\qquad G(x) = L + \ell(x - Q) + G(Q), \qquad x \ge S.$

The following proposition provides independent verification (that is, without relying on results from Chapter 3) that (7.17) to (7.20) uniquely determine the cost function G.

Proposition 7.3 *Suppose that $G : [0, S] \to \mathbb{R}$ is twice continuously differentiable and further satisfies (7.17) to (7.19). Now extend $G(\cdot)$ to $[0, \infty)$ by taking $G(x) = G(S) + \ell(x - S)$, $x \ge S$. Then $G(\cdot)$ is indeed the cost function for the control band policy with parameters (q, Q, S).*

Proof Let $f : \mathbb{R} \to \mathbb{R}$ be defined as follows: $f(x) = G(0) + G'(0)x$ if $x < 0$; $f(x) = G(x)$ if $0 \le x \le S$, and $f(x) = G(S) + G'(S)(x - S)$ if $x > S$. Then f satisfies all the hypotheses of Proposition 7.1. Using (7.17) to (7.20) plus the specific structure of the control band policy, Proposition 7.1 then gives us the following for $0 \le x \le S$ and arbitrary $T > 0$:

(7.21) $\quad E_x\left[e^{-\lambda T} G(Z_T)\right] = G(x) - E_x\left[\int_0^T e^{-\lambda t} \kappa Z_t \, dt\right]$

$$- E_x\left[\sum_{n=0}^{N(T)} e^{-\lambda T_n} \phi(\xi_n)\right].$$

Of course, $\{G(Z_T), T \ge 0\}$ is uniformly bounded, so the left side of (7.21) vanishes as $T \to \infty$. Thus letting $T \to \infty$ in (7.21) establishes the desired conclusion for $0 \le x \le S$; it is immediate from the policy description given earlier that the cost function grows linearly with slope ℓ on $[S, \infty)$. $\qquad\square$

7.3 Optimal policy parameters

Continuing the discussion of control band policies, we now seek to determine parameters (q, Q, S) such that the following three *optimality conditions* are satisfied:

(7.22) $\qquad\qquad G'(q) = -k,$

(7.23) $G'(Q) = \ell,$

(7.24) $G'(S-) = \ell.$

To explain the rationale for these conditions, let us fix the parameters (q, Q, S) and call the corresponding control band policy our *candidate policy*. If the controller starts in state S, immediately enforces a jump down to level $x < S$, and thereafter follows the candidate policy, his expected total cost will be $L + \ell(S - x) + G(x)$. If the candidate policy is to be optimal, this expression must be maximized by taking $x = Q$, which implies (7.23). In exactly the same way, by considering the various points $x > 0$ to which the controller could jump from level zero, we deduce (7.22) as a necessary condition for optimality. Condition (7.24) is a variant of the "principle of smooth fit" that appeared in Chapter 5, requiring that $G'(\cdot)$ be continuous on $[0, \infty)$, and it is necessary for optimality, as follows: if $G'(S-) < \ell$, then for all initial states $x < S$ that are sufficiently close to S, the candidate policy can be improved by enforcing an immediate downward jump to level Q, rather than waiting for Z to hit level S; and if $G'(S-) > \ell$, then for starting state S the candidate policy can be improved by *not* taking any immediate action. The first of those two assertions is relatively easy to prove, and readers are asked to provide details in Problem 7.1. The second one requires a more elaborate argument; see Problem 7.2.

To repeat, these arguments suggest that (7.22) to (7.24) are necessary for optimality. In the remainder of this section it will be shown that (7.22) to (7.24), together with (7.17) to (7.19), uniquely determine the policy parameters (q, Q, S), and then we will rigorously verify that the corresponding control band policy is optimal among *all* feasible policies, not just among control band policies. For that purpose it will be convenient to let

$$\pi(x) = \kappa/\lambda - G'(x), \qquad x \geq 0.$$

Differentiating (7.17) with respect to x gives

(7.25) $\Gamma\pi(x) - \lambda\pi(x) = 0, \qquad 0 \leq x \leq S,$

and (7.22) to (7.24) can be restated in terms of π as follows:

(7.26) $\pi(q) = c \quad \text{and} \quad \pi(Q) = \pi(S) = r,$

where

(7.27) $c := \kappa/\lambda + k \quad \text{and} \quad r := \kappa/\lambda - \ell.$

Using (7.1) we then have

(7.28) $0 < r < c < \infty.$

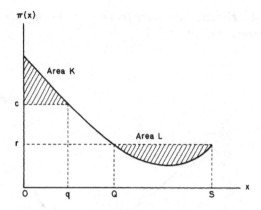

Figure 7.2 Optimal policy parameters.

To review, (7.25) restates our differential equation (7.17) for the cost function G in terms of π, while the three equalities in (7.26) restate our optimality conditions (7.22) to (7.24) in terms of π. The corresponding restatement of our boundary conditions (7.18) and (7.19) in terms of π is as follows:

$$(7.29) \qquad \int_0^q [\pi(x) - c]\, dx = K,$$

$$(7.30) \qquad \int_Q^S [r - \pi(x)]\, dx = L.$$

Figure 7.2 summarizes the relationships (7.26), (7.29), and (7.30) that will determine the optimal policy parameters (q, Q, S). Our construction involves a parametric family of functions $f_s(x)$ defined as follows: for $s > 0$ let

$$a_1(s) := (1 - e^{-\alpha_2 s})(e^{\alpha_1 s} - e^{-\alpha_2 s}) > 0,$$
$$a_2(s) := (e^{\alpha_1 s} - 1)/(e^{\alpha_1 s} - e^{-\alpha_2 s}) > 0,$$

and

$$(7.31) \qquad f_s(y) := r\left[a_1(s)e^{\alpha_1 y} + a_2(s)e^{-\alpha_2 y}\right], \qquad y \in \mathbb{R},$$

where $\alpha_1 := \alpha_1(\lambda)$ and $\alpha_2 := \alpha_2(\lambda)$ are defined as in Section 3.3. For any $s > 0$ it is easy to verify that $\Gamma f_s(x) - \lambda f_s(x) = 0, x \in \mathbb{R}$, with $f_s(0) = f_s(s) = r$. (In fact, $f_s(\cdot)$ *uniquely* satisfies the indicated differential equation subject to those boundary conditions, but we do not make use of that fact.) It is easy to verify that $f_s''(\cdot) > 0$, and hence we have the following.

Proposition 7.4   *For each $s > 0$ the function $f_s(\cdot)$ is strictly convex on $\mathbb{R}$ and has a minimum in $(0, s)$.*

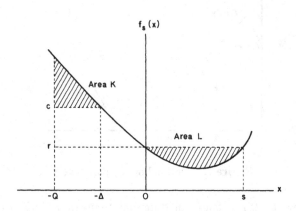

Figure 7.3 Determining the optimal parameters.

The function $f_s(\cdot)$ is pictured in Figure 7.3, whose other features will be explained shortly. Our plan is to construct a solution π of (7.25) and (7.26) by taking

(7.32) $\pi(x) = f_s(x - Q), \quad 0 \le x \le S, \quad \text{where } s = S - Q.$

To satisfy (7.30) it will then be necessary that

$$L = \int_Q^S [r - f_s(x - Q)] \, dx = \int_0^s [r - f_s(y)] \, dy$$

(7.33)

$$= rs - r\left(\frac{1}{\alpha_1} + \frac{1}{\alpha_2}\right)(e^{-\alpha_1 s} - 1)(1 - e^{-\alpha_2 s})/(e^{\alpha_1 s} - e^{-\alpha_2 s}).$$

The proof of the following is left as an exercise.

Proposition 7.5   *The right side of (7.33) increases strictly and continuously from 0 to ∞ as s increases from 0 to ∞, so there exists a unique $s > 0$ satisfying (7.33).*

Defining $\pi(\cdot)$ by (7.32) guarantees that the conditions $\pi(Q) = \pi(S) = r$ in (7.26) are satisfied. Also, the companion condition $\pi(q) = c$ will hold if and only if we take $q = Q - \Delta$ where

(7.34) $f_s(-\Delta) = c,$

as indicated in Figure 7.3. Finally, to determine Q we re-express (7.29) as

(7.35) $$\int_{-Q}^{-\Delta} [f_s(y) - c] \, dy = K.$$

With $s > 0$ and $\Delta > 0$ already fixed, it follows from Proposition 7.4 that (7.35) uniquely determines Q; this relationship is portrayed graphically in Figure 7.3. Using the definition (7.31) of $f_s(\cdot)$, we can evaluate the integral on the left side of (7.35) explicitly, which leads to the following restatement of (7.35):

(7.36) $$r \left[\frac{a_1(s)}{\alpha_1} (e^{-\alpha_1 \Delta} - e^{-\alpha_1 Q}) + \frac{a_2(s)}{\alpha_2} (e^{\alpha_2 Q} - e^{\alpha_2 \Delta}) \right] - c(Q - \Delta) = K.$$

Proposition 7.6 *Let $s > 0$, $\Delta > 0$, and $Q > \Delta$ be chosen sequentially to satisfy (7.33), (7.34), and (7.35), respectively, and then set $q := Q - \Delta$ and $S := Q + s$. The control band policy with parameters (q, Q, S) is optimal for the discounted impulse control problem stated in Section 7.1.*

Proof Our first task is to construct the cost function G for the (q, Q, S) control band policy by reversing the logic used earlier in this section. With s, Δ, Q, q, and S chosen in the indicated way, we define $f_s(\cdot)$ via (7.31) and $\pi(\cdot)$ via (7.32). Now let

(7.37) $$G(0) := \frac{\mu \kappa}{\lambda^2} - \frac{1}{\lambda} \left[\frac{1}{2} \sigma^2 \pi'(0) + \mu \pi(0) \right],$$

(7.38) $$G(x) := G(0) + \int_0^x [\kappa/\lambda - \pi(y)] \, dy, \qquad 0 \le x \le S,$$

and

(7.39) $$G(x) := G(S) + \ell(x - S), \qquad x > S.$$

Equation (7.37) sets the value of $G(0)$ so that $G(\cdot)$, as defined by (7.38), satisfies $(\lambda G - \Gamma G)(0+) = 0$. Because $\pi(\cdot)$ satisfies $(\lambda \pi - \Gamma \pi)(\cdot) = 0$ by construction, (7.38) ensures that $(\lambda G - \Gamma G)(x) = \kappa x$ for $0 \le x \le S$. Finally, because $\pi(\cdot)$ satisfies (7.29) and (7.30) by construction, we have that $G(\cdot)$ satisfies (7.18) and (7.19). Thus $G(\cdot)$ satisfies all the hypotheses of Proposition 7.3, implying that it is indeed the cost function for our control band policy with parameters (q, Q, S).

Our next task is to show that G satisfies the hypotheses of Proposition 7.2 and hence provides a lower bound for the cost function of any other feasible policy. First note that G is continuously differentiable on $[0, \infty)$

because $G'(S+) = \ell$ by (7.39) and $G'(S-) = \kappa/\lambda - \pi(S-) = \kappa/\lambda - f_s(s-) = \kappa/\lambda - r = \ell$ by construction. Next, it must be established that

$$(7.40) \qquad (\lambda G - \Gamma G)(x) \le \kappa x \qquad \text{for all } x \ge 0.$$

Of course, (7.40) holds with equality on $[0, S]$. As x increases toward S and then exceeds it, both $G(x)$ and $G'(x)$ are continuous, while $G''(\cdot)$ jumps upward from $G''(S-) = -\pi'(S-) = -f_s(s-) < 0$ (see Figure 7.3) to $G''(x) = 0$ for $x > S$. Thus $(\lambda G - \Gamma G)(S+) < \kappa S$. Finally, ΓG is constant to the right of S, while G itself is increasing linearly, so $(\lambda G - \Gamma G)(x) - \kappa x$ becomes even more negative as x increases from S. Thus (7.40) is confirmed. To verify the remaining hypotheses (7.10) and (7.11) of Proposition 7.2 one needs little more than the picture of $\pi(x) := G'(x) - \kappa/x$ given in Figure 7.2; we leave this as an exercise. $\qquad \square$

7.4 Impulse control with average cost criterion

Turning now to the undiscounted analog of the problem described in Section 7.1, we define a control policy $\{(T_n, \xi_n)\}$ exactly as before. Given a feasible policy $\{(T_n, \xi_n)\}$ with associated contents process Z, and given an initial state $x \ge 0$, we define the corresponding long-run average cost rate (or simply *average cost*) as

$$(7.41) \qquad \gamma(x) := \limsup_{T \to \infty} \frac{1}{T} E_x \left[\int_0^T \kappa Z_t \, dt + \sum_{n=0}^{N(T)} \phi(\xi_n) \right],$$

which is finite by (7.3) and (7.4). For a control band policy (see Section 7.2) it will be shown in the next section that the average cost γ exists as a limit (not just as a lim sup), is finite, is independent of the initial state x, and is the solution of a certain analytical problem. Later, in Section 7.6, we show how to choose the parameters (q, Q, S) of a control band policy so that its average cost γ is minimal over all feasible policies, regardless of the initial state x. The following analog of Proposition 7.2 will be used to prove the optimality of that specific control band policy.

Proposition 7.7 *Let $\gamma(\cdot)$ be the average cost for an arbitrary feasible policy, defined by (7.41). Suppose that $f : [0, \infty) \to \mathbb{R}$ is C^1, has a bounded derivative, is piecewise C^2, and further satisfies the following three conditions:*

$$(7.42) \qquad \Gamma f(0) - \Gamma f(x) \le \kappa x \qquad \text{for almost all } x \ge 0,$$

(7.43) $f(x) - f(y) \le K + k(y - x)$ *for* $0 \le x < y,$

(7.44) $f(x) - f(y) \le L - \ell(y - x)$ *for* $0 \le y < x.$

Then $\Gamma f(0) \le \gamma(x)$ *for all* $x \ge 0.$

Proof Let Z be the contents process under a feasible policy. We proceed exactly as in the proof of Proposition 7.2, first observing that (7.43) and (7.44) imply $f(x) - f(y) \le \phi(x - y)$, and hence that

(7.45) $-\theta_n \le \phi(\xi_n)$ for all $n = 0, 1, \ldots,$

where θ_n is defined by (7.7). We then apply Proposition 7.1 with $\lambda = 0$ to get

(7.46) $E_x[f(Z_T)] = f(x) + E_x\left[\int_0^T \Gamma f(Z_t)\, dt\right] + E_x\left[\sum_{n=0}^{N(T)} \theta_n\right]$

for arbitrary $x \ge 0$ and $T > 0$. Substituting (7.42) and (7.45) into (7.46) and rearranging terms, we have

(7.47) $f(x) \le E_x\left\{\int_0^T [\kappa Z_t - \Gamma f(0)]\, dt\right\} + E_x\left[\sum_{n=0}^{N(T)} \phi(\xi_n)\right] + E_x[f(Z_T)].$

We now divide both sides of (7.47) by T and take the lim sup as $T \to \infty$. The left side obviously vanishes, and it follows from (7.4) that the last term on the right side vanishes as well. Thus, using the definition (7.41), we have from (7.47) that $0 \le \gamma(x) - \Gamma f(0)$, as desired.   □

7.5 Relative cost functions

The function g appearing in the following proposition is called the *relative cost function* for a given control band policy. Its interpretation, and its relationship to quantities introduced earlier in the discounted formulation, will be discussed after the proposition has been proved.

Proposition 7.8 *Let* (q, Q, S) *be the parameters of a control band policy. There exists a function* $g : [0, S] \to \mathbb{R}$, *unique up to an additive constant, that is twice continuously differentiable and satisfies the following three conditions:*

(7.48) $\Gamma g(x) = \Gamma g(0) - \kappa x,$ $0 \le x \le S,$

(7.49) $g(0) = g(q) + K + kq,$

(7.50) $g(S) = g(Q) + L + \ell(S - Q).$

Furthermore, the average cost γ for the control band policy exists as a limit and is given by $\gamma = \Gamma g(0)$, *independent of the initial state.*

Proof   We address only the case $\mu \neq 0$; the case $\mu = 0$ is treated similarly. The general solution of (7.48) is

$$(7.51) \qquad g(x) = Ax + Be^{-2\mu x/\sigma^2} - \frac{\kappa}{2\mu}x^2 + C,$$

where A, B, and C are constants. In our case C can be chosen arbitrarily, and then A and B are uniquely determined by the boundary conditions (7.49) and (7.50).

To prove the last statement of the proposition, we apply Proposition 7.1 with $\lambda = 0$ and the function f defined as follows: $f(x) = g(x)$ for $0 \leq x \leq S$ and $f(x) = g(S) + g'(S)(x - S)$ for $x > S$. Proceeding exactly as in the proofs of Proposition 7.2 and Proposition 7.7, we then deduce from (7.48) to (7.50) that, for $0 \leq x \leq S$ and arbitrary $T > 0$,

$$(7.52) \quad E_x\left[g(Z_T)\right] = g(x) - E_x\left\{\int_0^T [\kappa Z_t - \Gamma g(0)]\, dt\right\} - E_x\left[\sum_{n=0}^{N(T)} \phi(\xi_n)\right].$$

Because $\{g(Z_T),\ T \geq 0\}$ is bounded, we can then divide both sides of (7.52) by T and let $T \rightarrow \infty$ to conclude the following: the lim sup on the right side of (7.41) exists as a limit and equals $\Gamma g(0)$, independent of the initial state x, which is the desired conclusion. Finally, for an initial state $x > S$, the expected total cost over an interval $[0, T]$ is $\ell(x - S)$ plus the expected total cost over that same interval with initial state $x = S$, and thus we reach the same conclusion.   □

For an interpretation of $g(\cdot)$, let us assume $0 \leq x \leq S$ and denote by $C(x, T)$ the expected total cost over $[0, T]$ when following the control band policy. Defining $\gamma := \Gamma g(0)$, the right side of (7.52) is then $g(x) + \gamma T - C(x, T)$. Also, using renewal theory (that is, using the regenerative structure of the process Z) as in Section 6.5, it can be shown that the left side of (7.52) approaches a constant not depending on x as $T \rightarrow \infty$, because Z_T converges in distribution as $T \rightarrow \infty$. Thus we have from (7.52) that

$$(7.53) \qquad C(x, T) - C(y, T) \longrightarrow g(x) - g(y) \qquad \text{as } T \rightarrow \infty$$

for $0 \leq x, y \leq S$. In words, $g(\cdot)$ expresses the relative costliness of different initial states over a long planning horizon, which suggests the following natural extension:

$$g(x) = g(S) + \ell(x - S) \qquad \text{for } x > S.$$

To elaborate further on the interpretation of $g(\cdot)$, let us fix the parameters (q, Q, S) of a control band policy and denote by $G_\lambda(x)$ the expected present value of total costs incurred over the infinite horizon when the starting state is $x \geq 0$ and the interest rate is $\lambda > 0$; this augments in an obvious way the notation used earlier in Section 7.1. Using the regenerative structure of a control band policy, it is quite easy to show that

$$(7.54) \qquad \lambda G_\lambda(x) \longrightarrow \gamma \qquad \text{as } \lambda \downarrow 0 \quad \text{for all } x \geq 0$$

(this is a *Tauberian theorem*), and that

$$(7.55) \qquad G_\lambda(x) - G_\lambda(y) \longrightarrow g(x) - g(y) \qquad \text{as } \lambda \downarrow 0$$

for any $x, y \geq 0$, where γ and $g(\cdot)$ are as specified in Proposition 7.8.

Suppose for the sake of concreteness that we fix the additive constant in the definition of g (see Proposition 7.8) by requiring that $g(0) = 0$ and also define $g_\lambda(x) := G_\lambda(x) - G_\lambda(0)$ for $\lambda > 0$ and $x \geq 0$. Then (7.55) gives

$$(7.56) \qquad g_\lambda(x) \longrightarrow g(x) \qquad \text{as } \lambda \downarrow 0,$$

and the differential equation (7.17) that characterizes $G_\lambda(\cdot)$ can be restated in terms of $g_\lambda(\cdot)$ as follows:

$$(7.57) \qquad \lambda g_\lambda(x) + \lambda G_\lambda(0) - \Gamma g_\lambda(x) = \kappa x, \qquad 0 \leq x \leq S.$$

The first term on the left side of (7.57) vanishes as $\lambda \downarrow 0$ by (7.56), and the second term converges to $\Gamma g(0)$ by (7.54) and Proposition 7.8. Thus, if we send $\lambda \downarrow 0$ in (7.57) and formally interchange limits and derivatives in the third term on the left, we arrive at the differential equation (7.48) that characterizes $g(\cdot)$ in Proposition 7.8. Furthermore, the boundary conditions (7.18) and (7.19) for $G_\lambda(\cdot)$ can be equivalently stated with $g_\lambda(\cdot)$ in place of $G_\lambda(\cdot)$, and if we formally let $\lambda \downarrow 0$ in those restated boundary conditions we get (7.49) and (7.50). Thus the analytical characterization of $g(\cdot)$ via (7.48) to (7.50) is precisely what one expects by taking formal limits in (7.17) to (7.19).

7.6 Average cost optimality

Let us consider the optimality conditions (7.22) to (7.24) for the discounted impulse control problem, formally letting $\lambda \downarrow 0$ in those equations. Reasoning as in the previous paragraph, using (7.57) and (7.56), we arrive at the following *average cost optimality conditions* for a control band policy:

$$(7.58) \qquad g'(q) = -k,$$

(7.59)
$$g'(Q) = \ell,$$

(7.60)
$$g'(S) = \ell.$$

In this section we show how to contruct parameters (q, Q, S) for a control band policy that satisfy (7.58) to (7.60), then rigorously verify that the resulting policy is average cost optimal.

The development now diverges somewhat from that in Section 7.3. In the current context we define

$$\pi(x) := \ell - g'(x) \qquad \text{for } 0 \le x \le S.$$

Differentiating (7.48) with respect to x gives

(7.61)
$$\Gamma\pi(x) = \kappa \qquad \text{for } 0 \le x \le S,$$

and we restate (7.58) to (7.60) as follows:

(7.62)
$$\pi(q) = k + \ell,$$

(7.63)
$$\pi(Q) = \pi(S) = 0.$$

Finally, we restate our boundary conditions (7.49) and (7.50) in terms of π as follows:

(7.64)
$$\int_0^q [\pi(x) - (k + \ell)] \, dx = K,$$

(7.65)
$$-\int_Q^S \pi(x) \, dx = L.$$

As a parallel to the development in Section 7.3, we now show that (7.61)–(7.65) uniquely determine both $\pi(\cdot)$ and the policy parameters (q, Q, S). More specifically, a solution of these relationships will be constructed from the parametric family of functions $f_s(\cdot)$ identified in the following proposition.

Proposition 7.9 *For each $s > 0$, there exists a unique function $f_s(\cdot)$ on $\mathbb{R}$ that satisfies*

(7.66)
$$\Gamma f_s(x) = \kappa \qquad \text{for all } x \in \mathbb{R}$$

and

(7.67)
$$f_s(0) = f_s(s) = 0.$$

Furthermore, $f_s(\cdot)$ is strictly convex and the functional

(7.68)
$$I(s) := -\int_0^s f_s(x) \, dx$$

increases strictly and continuously from 0 to ∞ as s increases from 0 to ∞.

Proof If $\mu = 0$, then the unique solution of (7.66) and (7.67) is $f_s(x) = x(x - s)/\sigma^2$, and the remaining claims are obviously true. If $\mu \neq 0$, then the unique solution is

$$f_s(x) = \frac{\kappa}{\mu}x - \frac{\kappa}{\mu}s\left(\frac{1 - e^{-2\mu x/\sigma^2}}{1 - e^{-2\mu s/\sigma^2}}\right).$$

Thus $f_s''(\cdot) > 0$ on all of $\mathbb{R}$, implying strict convexity. Verification of the final claim is left as an exercise. □

The following derivation of optimal policy parameters (q, Q, S) parallels precisely the derivation of discount-optimal parameters in Section 7.3. In particular, Figure 7.3, originally intended as a graphical representation of the discount-optimal computations, applies equally well in the current context if we redefine $r := 0$ and $c := k + \ell$.

As a first step, let $s > 0$ be defined by the requirement that

(7.69) $$I(s) = L,$$

where $I(\cdot)$ is defined by (7.68). Then $\Delta > 0$ and $Q > \Delta$ are defined by

(7.70) $$f_s(-\Delta) = k + \ell$$

and

(7.71) $$\int_{-Q}^{-\Delta} [f_s(x) - (k + \ell)]\, dy = K.$$

It is immediate from Proposition 7.9 that s, Δ, and Q are uniquely determined by (7.69), (7.70), and (7.71), respectively. Finally, let

(7.72) $$q := Q - \Delta$$

and

(7.73) $$S := Q + s.$$

Proposition 7.10 *If the parameters (q, Q, S) are chosen to satisfy (7.69) to (7.73), where $f_s(\cdot)$ is the function described in Proposition 7.9 and $I(\cdot)$ is defined by (7.68), then the corresponding control band policy is average cost optimal.*

Remark Because an explicit formula for $f_s(\cdot)$ is provided in the proof of Proposition 7.9 (actually two different formulas are provided, depending on whether $\mu = 0$ or $\mu \neq 0$), the integrals in (7.69) and (7.71) can also be expressed as explicit formulas; see Theorem 1 and Theorem 2 of Ormeci et al. (2008).

Proof The proof is almost identical to that of Proposition 7.6, so it will only be sketched. Let

$$(7.74) \qquad \pi(x) := f_s(x - Q) \qquad\qquad \text{for } 0 \leq x \leq S,$$

$$(7.75) \qquad g(x) := \int_0^x [\ell - \pi(y)] \, dy \qquad \text{for } 0 \leq x \leq S,$$

and

$$(7.76) \qquad g(x) := g(S) + \ell(x - S) \qquad\qquad \text{for } x > S.$$

By construction, π satisfies (7.61), (7.64), and (7.71), so g satisfies (7.48) to (7.50). Thus, by Proposition 7.8, g is the relative cost function for the control band policy with parameters (q, Q, S), and $\Gamma g(0)$ is the average cost for that policy. In addition, because π further satisfies (7.62) and (7.63), the function $f = g$ satisfies all the hypotheses of Proposition 7.7, so the average cost function $\gamma(x)$ for an arbitrary feasible policy is bounded below by $\Gamma g(0)$.   □

7.7 Instantaneous control with discounting

Let us consider now the control problem described in Section 7.1, altered by the assumption that $K = L = 0$. The letter L, no longer needed to denote a fixed cost of control, will be re-used in this section and the next one with a different meaning.

In the current context a *control policy* (or just *policy*) is defined as a pair of processes L and U such that

L and U are adapted, increasing, and positive.

Interpret L_t as the cumulative increase in system contents effected by the controller up to time t, and U_t as the corresponding cumulative decrease effected. The letters L and U were used in Chapter 6 to denote increasing processes associated with the *lower* and the *upper* boundaries, respectively. In the current context, this notation foretells the form of the optimal policy. We associate with policy (L, U) the *controlled process* $Z := X + L - U$. A *feasible* policy in the current context is one that satisfies (7.2), (7.4), and the following replacement for (7.3):

$$\sup_{t \geq 1} \frac{1}{t} E_x(L_t + U_t) < \infty \qquad \text{for all } x \geq 0.$$

To simplify discounted cost expressions, let us define

$$\int_0^\infty e^{-\lambda t} \, dL := L_0 + \int_{(0,\infty)} e^{-\lambda t} \, dL$$

and similarly with dU in place of dL. This notational convention will be employed throughout this section and the next one without further comment. We associate with a feasible policy (L, U) the *cost function*

$$G(x) := E_x \left\{ \int_0^\infty e^{-\lambda t} [\kappa Z_t \, dt + k \, dL + \ell \, dU] \right\}, \qquad x \geq 0,$$

and (L, U) is said to be *optimal* if $G(x)$ is minimal (among the cost functions associated with feasible policies) for each $x \geq 0$.

Recall that the optimal impulse control policy derived in Section 7.3 enforces an upward jump of size q when the contents process Z hits zero and enforces a downward jump of size $s := S - Q$ when Z hits S. It is intuitively obvious, and easy to check mathematically, that the optimal values of q and s vanish as the fixed control costs approach zero. Moreover, the behavior of the contents process Z approaches that of reflected Brownian motion on the interval $[0, S]$. Thus one is led to conjecture that there exists an optimal policy of the following form.

Given a policy parameter $S > 0$ and an initial state $x \in [0, S]$, we define a *barrier policy* as the pair of processes L and U that enforce a lower reflecting barrier at zero and an upper reflecting barrier at S (see Section 6.1); if $X_0 = x > S$, we take $U_0 = x - S$, and then future increments of U are defined in the obvious way. From Corollary 6.3 we know that the cost function G for a barrier policy satisfies

(7.77) $$\lambda G(x) - \Gamma G(x) = \kappa x, \qquad 0 \leq x \leq S,$$

(7.78) $$G'(0) = -k \quad \text{and} \quad G'(S) = \ell,$$

to which we add

(7.79) $$G(x) = G(S) + \ell(x - S), \qquad x > S.$$

Proposition 6.4 shows how to construct an explicit solution for (7.77) and (7.78) given an arbitrary $S > 0$, but the explicit solution is not needed for current purposes.

Seeking now to optimize the policy parameter S, it is natural to "take a limit" as the fixed control costs approach zero in the optimality conditions (7.22) to (7.24) for our impulse control problem. Because $q \downarrow 0$ as the fixed control costs vanish, the first and last of those optimality conditions simply reduce to our current boundary conditions (7.78). However, (7.23) and (7.24) together imply that, when the parameters (q, Q, S) of a control band policy are optimized in the impulse control problem, one has $G''(x) = 0$ for some $x \in [Q, S]$. As noted earlier, $s := S - Q \downarrow 0$ as fixed control costs vanish (that is, the optimal policy parameters Q and S coalesce in the limit), which suggests the following optimality condition for our instantaneous control problem:

$$(7.80) \qquad\qquad G''(S-) = 0.$$

Of course, (7.79) implies $G''(x) = 0$ for all $x > S$, so (7.80) is equivalent to requiring that $G''(\cdot)$ be continuous on $[0, \infty)$. In the remainder of this section it will be shown that (7.80) uniquely determines S, and that the resulting barrier policy is optimal. Problems 7.3 and 7.4 sketch a direct argument, not involving a limit as fixed control costs vanish, that (7.80) is *necessary* for optimality.

The remaining argumentation parallels that in Section 7.3 but is simpler in every respect, so many details will be omitted. To begin, let $\pi(x) := \kappa/\lambda - G'(x)$, $c := \kappa/\lambda + k$, and $r := \kappa/\lambda - \ell$ as before, so that $0 < r < c < \infty$ by (7.1). Then (7.77), (7.78), and (7.80) can be restated in terms of π as follows:

$$(7.81) \qquad\qquad \lambda\pi(x) - \Gamma\pi(x) = 0, \qquad 0 \le x \le S,$$
$$(7.82) \qquad\qquad \pi(0) = c \quad \text{and} \quad \pi(S) = r,$$
$$(7.83) \qquad\qquad \pi'(S) = 0.$$

In place of the parametric family $f_s(\cdot)$ used in Section 7.3, we now define the single function

$$f(x) = \left(\frac{r}{\alpha_1 + \alpha_2}\right)(\alpha_2 e^{-\alpha_1 x} + \alpha_1 e^{\alpha_2 x}), \qquad x \in \mathbb{R},$$

which satisfies $\lambda f(x) - \Gamma f(x) = 0$ for all $x \in \mathbb{R}$, $f(0) = r$, and $f'(0) = 0$. Then $f(x)$ is strictly convex on $\mathbb{R}$ with a minimum at $x = 0$, and $f(x) \to \infty$ as $|x| \to \infty$, from which the following is immediate.

Proposition 7.11 *There exists a unique $S > 0$ such that $f(-S) = c$.*

To obtain a solution of (7.81) to (7.83), let $\pi(x) := f(x - S)$, $0 \le x \le S$, then define G in terms of π via (7.37) to (7.39), exactly as in Section 7.3. The resulting function G is twice continuously differentiable and convex on $[0, \infty)$, and it satisfies (7.77) to (7.80). From this it is immediate that

$$(7.84) \qquad\qquad -k \le G'(x) \le \ell, \qquad x \ge 0,$$

and for reasons to be explained, we also have

$$(7.85) \qquad\qquad \lambda G(x) - \Gamma G(x) \le \kappa x, \qquad x \ge 0.$$

Of course, (7.77) says that (7.85) holds with equality on $[0, S]$, and from (7.78) to (7.80) we know that $\Gamma G(x) = \Gamma G(S)$ for $x \ge S$. Thus (7.85) asserts simply that $\lambda \ell (x - S) \le \kappa (x - S)$ for $x \ge S$, which is true by (7.1).

Proposition 7.12 *If S is chosen as in Proposition 7.11, then the corresponding barrier policy is optimal.*

Proof Let $x \ge 0$ be arbitrary, let G be the cost function for the specified barrier policy as in the preceding paragraph, and let $\tilde{G}$ be the cost function for another feasible policy (L, U) with contents process Z. We first show that $G(x) \le \tilde{G}(x)$ under the assumption that L and U are both continuous processes, indicating afterward the changes that are needed if L and U have jumps. Because G has a bounded derivative we can argue exactly as in the proof of Proposition 7.1, using Itô's formula and integration by parts, that

$$(7.86) \quad E_x\left[e^{-\lambda T} G(Z_T)\right] = G(x) + E_x\left[\int_0^T e^{-\lambda t}(\Gamma G - \lambda G)(Z_t)\,dt\right]$$
$$+ E_x\left[\int_0^T e^{-\lambda t} G'(Z_t)\,d(L - U)\right]$$

for any $T > 0$. From (7.84) we have that $G'(Z_t)\,d(L - U) \ge -k\,dL - \ell\,dU$. Substituting this inequality and (7.85) into (7.86), we have the following after rearranging terms:

$$(7.87) \quad G(x) \le E_x\left\{\int_0^T e^{-\lambda t}[\kappa Z_t\,dt + k\,dL + \ell\,dU]\right\} + E_x\left[e^{-\lambda T} G(Z_T)\right].$$

It follows from (7.4) and the continuity of G that the last term on the right side of (7.87) vanishes as $T \to \infty$, and the first term converges to $\tilde{G}(x)$. Thus we have $G(x) \le \tilde{G}(x)$ as desired.

Considering now the case where L and U may contain jumps, let us denote by ΔL_t, ΔU_t, and $\Delta G(Z)_t$ the jump in L, the jump in U, and the jump in $G(Z)$, respectively, at time t. We then have from (7.84) that

$$\Delta G(Z)_t \leq k\Delta L_t + \ell\Delta U_t.$$

Thus, when we apply Proposition 4.15 to the process $\exp(-\lambda T)G(Z_T)$, as in the proof of Proposition 7.1, we eventually arrive at the same inequality (7.87), and hence at the same conclusion, $G(x) \leq \tilde{G}(x)$. □

7.8 Instantaneous control with average cost criterion

Finally, we consider the instantaneous control problem of Section 7.7 (that is, we assume $K = L = 0$) without discounting. The analysis involves a sequence of propositions that are analogous to, but simpler than, results derived earlier for either the impulse control problem with average cost criterion (Sections 7.4 to 7.6) or else the instantaneous control problem with discounting (Section 7.7). In the interest of brevity, proofs will merely be sketched, with details left to the reader. •

A feasible policy (L, U) and the associated contents process Z are defined exactly as in Section 7.7, and following the pattern established in Section 7.4, we define the associated average cost function

$$(7.88) \qquad \gamma(x) := \limsup_{T\to\infty} \frac{1}{T}E_x\left\{\int_0^T [\kappa Z_t\,dt + k\,dL + \ell\,dU]\right\}$$

for $x \geq 0$. Also defined as in Section 7.7 is a barrier policy with parameter $S > 0$. In the obvious way, a feasible policy (L, U) is said to be optimal if its average cost $\gamma(x)$ is minimal, among all feasible policies, for all $x \geq 0$.

Proposition 7.13 *Let $S > 0$ be fixed but arbitrary. There exists a function $g : [0, S] \to \mathbb{R}$, unique up to an additive constant, that is twice continuously differentiable and satisfies the following three conditions:*

$$\Gamma g(x) = \Gamma g(0) - \kappa x, \qquad 0 \leq x \leq S,$$
$$g'(0) = -k \quad and \quad g'(S) = \ell.$$

Furthermore, the average cost γ for the barrier policy with parameter S is $\gamma = \Gamma g(0)$, independent of the initial state x.

Proof Exactly as in the proof of Proposition 7.8, the general solution of the indicated differential equation is given by (7.51); the constant C can be chosen arbitrarily, and the constants A and B are uniquely determined by the boundary conditions $g'(0) = -k$ and $g'(S) = \ell$. To prove the last statement we proceed exactly as in the proof of Proposition 7.8, except that now the final term on the right side of (7.52) is

$$-E_x\left\{\int_0^T [k\,dL + \ell\,dU]\right\}.$$

Dividing both sides of (7.52) by T and letting $T \to \infty$ then gives $\gamma = \Gamma g(0)$, as desired. $\qquad\square$

As in Section 7.5, the function g described in the preceding proposition is called the relative cost function for the barrier policy with parameter S, and we extend it to all of $[0, \infty)$ by setting $g(x) = g(S) + \ell(x - S)$ for $x > S$. Extrapolating from optimality conditions derived in Section 7.6 and Section 7.7, one naturally conjectures that the barrier policy with parameter S is optimal in the current context if and only if its relative cost function g satisfies

(7.89) $$g''(S-) = 0;$$

because $g''(S+) = 0$ as a matter of definition, (7.89) is equivalent to the requirement that g have a continuous second derivative on all of $[0, \infty)$.

Defining $\pi(x) := \ell - g'(x)$ for $0 \le x \le S$ as in Section 7.6, we now seek a solution of the following problem, which is analogous to (7.61) to (7.65) but simpler: find $S > 0$ and a twice continuously differentiable function π on $[0, S]$ such that

(7.90) $$\Gamma\pi(x) = \kappa \qquad \text{for } 0 \le x \le S,$$

(7.91) $$\pi(0) = k + \ell \quad \text{and} \quad \pi(S) = 0,$$

(7.92) $$\pi'(S) = 0.$$

Proposition 7.14 *If $\mu = 0$, set $f(x) := \frac{\kappa}{\sigma^2}x^2$ for all $x \in \mathbb{R}$, and otherwise set*

$$f(x) := \frac{\kappa}{\mu}x + 1 - e^{-2\mu x/\sigma^2}, \qquad x \in \mathbb{R}.$$

Let $S > 0$ be the unique solution of the equation $f(-S) = k + \ell$ and define $\pi(x) := f(x - S)$. Then S and $\pi(\cdot)$ jointly satisfy (7.90) to (7.92).

Proof This is just a matter of verification, which is left as an exercise. $\quad\square$

Proposition 7.15 *Let $S > 0$ be chosen as in Proposition 7.14. The relative cost function g for the barrier policy with parameter S (see Proposition*

7.13) is twice continuously differentiable on $[0, \infty)$ *and satisfies*

(7.93) $\Gamma g(x) \geq 0 \geq \Gamma g(0) - \kappa x$ *for all* $x \geq 0$,

(7.94) $-k \leq g'(x) \leq \ell$ *for all* $x \geq 0$.

Corollary 7.16   *The barrier policy with parameter S is optimal.*

Proof Proceeding exactly as in the proof of Proposition 7.10, we define g in terms of π on $[0, S]$ via (7.75), then extend g to all of $[0, \infty)$ via (7.76). Now π satisfies (7.90) to (7.92) by Proposition 7.14, from which it follows that g satisfies the conditions specified in Proposition 7.13, plus (7.89). Thus $g''(\cdot)$ is continuous on $[0, \infty)$, and Proposition 7.13 says that g is the relative cost function for the barrier policy with parameter S, whose average cost is $\gamma = \Gamma g(0)$, independent of the initial state x.

Proposition 7.14 provides an explicit formula for $\pi(\cdot)$ on $[0, S]$, from which it can be verified that $g'(\cdot)$ increases monotonically from $g'(0) = -k$ to $g'(S) = \ell$; of course, $g'(x) = \ell$ for all $x \geq \ell$ as a matter of definition. Thus g is convex on $[0, \infty)$ and satisfies (7.94). By construction, g satisfies (7.93) with equality on $[0, S]$, and (7.89) gives $\Gamma g(x) = \Gamma g(S)$ for all $x > S$, so (7.93) holds on $[S, \infty)$ as well. This concludes the proof of Proposition 7.15.

The proof of the corollary mimics that of Proposition 7.12 (except that $\lambda = 0$ in the current context), as follows. Let (L, U) be an arbitrary feasible policy, and denote by Z its contents process. Assume initially that L and U are both continuous. An application of Itô's formula gives the following analog of (7.86):

$$(7.95) \quad E_x[g(Z_T)] = g(x) + E_x\left[\int_0^T \Gamma g(Z_t)\, dt\right]$$
$$+ E_x\left[\int_0^T g'(Z_t)\,(dL - dU)\right]$$

for any $x \geq 0$ and $T > 0$. From (7.94) we have $g'(Z_t)\,(dL - dU) \geq -k\,dL - \ell\,dU$; substituting that inequality and (7.93) into (7.95) then rearranging terms, we have

$$g(x) + T\Gamma g(0) \leq E_x\left\{\int_0^T [\kappa Z_t\, dt + k\, dL + \ell\, dU]\right\} + E_x[g(Z_T)].$$

Now divide both sides by T and take the lim sup as $T \to \infty$. The second term on the right side vanishes, because g is continuous and a feasible policy is required to satisfy (7.4), and thus we arrive at the inequality

$\Gamma g(0) \leq \gamma(x)$, where $\gamma(\cdot)$ is defined by (7.88). This concludes the proof when L and U are both continuous, and the extension to allow jumps in the controls is accomplished as in the proof of Proposition 7.12.   □

7.9 Cash management

As an application of the instantaneous control problem with discounting (Section 7.7), let us consider the so-called *stochastic cash management problem*. Here Z_t represents the content of a cash fund into which various types of income or revenue are automatically channeled, and out of which operating disbursements are made. In our formulation, the net of such routine deposits less routine disbursements is modeled by a (μ, σ) Brownian motion. That is, in the absence of managerial intervention, the content of the fund fluctuates as a (μ, σ) Brownian motion X. Let us suppose that wealth not held as cash is invested in securities (hereafter called *bonds*) that pay interest continuously at rate λ. Denote by S_t the dollar value of bonds held at time t. Money can be transferred at any time from the cash fund to buy bonds, but a transaction cost of ℓ dollars must be paid for each dollar so transferred. That is, management gets only $1 - \ell$ dollars worth of bonds in exchange for one dollar of cash. Similarly, bonds can be sold at any time to obtain additional cash, but management must give up $1 + k$ dollars worth of bonds to obtain one dollar of cash. Management is obliged to keep the content of the cash fund positive, and the firm's initial wealth (cash plus bonds) is large enough that we can safely ignore the possibility of ruin.

Let U_t denote the cumulative amount of cash used to buy bonds up to time t, each dollar of which buys only $1 - \ell$ dollars worth of bonds. Similarly, let L_t denote the cumulative amount of cash generated by sale of bonds up to time t, each dollar of which requires liquidation of $1 + k$ dollars worth of bonds. The content of the cash fund is then $Z_t = X_t + L_t - U_t$ at time t, with $X_0 := Z_0$ by convention. The dynamics of the process S_t are given by

$$dS_t = \lambda S_t \, dt + (1 - \ell) \, dU_t - (1 + k) \, dL_t,$$

which means that

(7.96) $$S_T = S_0 e^{\lambda T} + \int_0^T e^{\lambda(T-t)} [(1 - \ell) \, dU - (1 + k) \, dL].$$

Let us suppose that management seeks to maximize the expected total wealth $E(Z_T + S_T)$ at some specified distant time T. This is equivalent to maximizing the expected present value

(7.97) $$e^{-\lambda T} E(S_T + Z_T).$$

It can be shown that $\exp(-\lambda T) E(Z_T)$ vanishes as $T \to \infty$ under any policy worthy of consideration (see Problem 7.5). Thus, substituting (7.96) into (7.97), sending $T \to \infty$, and ignoring the uncontrollable term S_0, we arrive at the objective of *maximizing*

(7.98) $$E\left\{ \int_0^\infty e^{-\lambda t} [(1 - \ell) \, dU - (1 + k) \, dL] \right\}.$$

Using integration by parts and the definition $Z := X + L - U$, it is easy to show that maximizing (7.98) is equivalent to minimizing

$$E\left\{ \int_0^\infty e^{-\lambda t} [\lambda Z_t \, dt + k \, dL + \ell \, dU] \right\},$$

which is a particular case of the instantaneous control problem treated in Section 7.7. Here the problem parameters k and ℓ represent actual out-of-pocket transaction costs, whereas the holding cost parameter $\kappa = \lambda$ reflects an *opportunity loss* on assets held as cash.

If, in addition to the proportional transaction costs already discussed, each conversion of cash to bonds involves a fixed fee of L dollars and each conversion of bonds to cash involves a fixed fee of K dollars, irrespective of the amounts being converted, then we have instead the impulse control problem introduced in Section 7.1.

7.10 Sources and literature

Sections 7.1 to 7.3 (impulse control with discounting) follow closely the paper by Harrison et al. (1983), and Sections 7.4 to 7.6 (impulse control with average cost criterion) are based on the paper by Ormeci et al. (2008). Egami (2008) considers optimal impulse control of a general one-dimensional diffusion process, developing an elegant general solution that parallels the findings of Dayanik and Karatzas (2003) on optimal stopping of a one-dimensional diffusion. A classical reference on impulse control is the book by Bensoussan and Lions (1975).

The instantaneous control problem with discounting that was studied in Section 7.7 was originally analyzed by Harrison and Taylor (1978); Harrison and Taksar (1983) treated a more general problem where the inventory holding cost is a convex, increasing function of the inventory level. In such problems the optimal controls (L, U) are continuous, but their points of increase form a set of (Lebesgue) measure zero. Control problems whose optimal solutions have this property are often called *singular* control problems. The first paper on singular control was by Bather and Chernoff (1967), but it was the work of Beneš et al. (1980) that launched the modern literature on this subject.

7.11 Problems and complements

Problem 7.1 As in Section 7.2, let $G(\cdot)$ be the cost function for a control band policy with parameters (q, Q, S). Now suppose that $G'(S-) < \ell$. Show that for all initial states x in an interval $(S-\epsilon, S)$, where $\epsilon > 0$, a cost strictly smaller than $G(x)$ can be achieved by immediately enforcing a downward jump to Q, and thereafter following the control band policy with parameters (q, Q, S).

Problem 7.2 (*Continuation*) Now consider the reverse situation where $G'(S-) > \ell$. Starting in state S, consider a policy that takes no action until the first time τ at which Z exits the interval $(S - \epsilon, S + \epsilon)$, reverting then to the control band policy. Use the result of Problem 4.17 to show that, for all $\epsilon > 0$ sufficiently small, the expected present value of the costs incurred under this policy is strictly smaller than $G(S)$.

Problem 7.3 In the setting of Section 7.7 (instantaneous control with discounting), consider a barrier policy with parameter $S > 0$, and suppose that $G''(S-) < 0$. Show that for all initial states x in an interval $(S-\epsilon, S)$, where $\epsilon > 0$, a cost strictly smaller than $G(x)$ can be achieved by immediately enforcing a downward jump to $S - \epsilon$, and thereafter following the barrier policy with parameter S.

Problem 7.4 (*Continuation*) Now consider the reverse situation where $G''(S-) > 0$. Show that for all initial states x in an interval $(S, S + \epsilon)$, a cost strictly smaller than $G(x)$ can be achieved as follows: enforce only a lower reflecting barrier at zero up until the contents process Z hits level $S + \epsilon$, then enforce a downward jump to level S and thereafter follow the barrier policy with parameter S.

Problem 7.5 To justify the claim made immediately after (7.97), flesh out the following skeletal argument, which involves the one-sided reflection mapping (see Section 2.2). First, regardless of how one chooses the monotone control U, one can maximize $S_t + Z_t$ almost surely for all $t \geq 0$

simultaneously by taking

(7.99) $$L_t = \sup_{0 \le s \le t} (X_s - U_s)^-, \qquad t \ge 0.$$

Thus, given our goal of maximizing $\exp(-\lambda T)E_x(S_T + Z_T)$ for some fixed T, the best choice of L to accompany any choice of U is (7.99).

Second, it follows from (7.99) that, for any $t \ge 0$,

(7.100) $$Z_t := X_t + L_t - U_t \le Z_t^*,$$

where

$$Z_t^* := X_t - \sup_{0 \le s \le t} X_s.$$

Finally, the process Z^* is (μ, σ) reflected Brownian motion, and it follows from formula (3.63) that

$$e^{-\lambda t} E_x(Z_t^*) \longrightarrow 0 \qquad \text{as } t \to \infty.$$

From (7.100) we then have $\exp(-\lambda t)E_x(Z_t) \to 0$ as well.

8

Brownian Models of Dynamic Inference

A commonly used model of "sequential learning" is the following. A decision maker is initially uncertain about a parameter θ, and in each period $t = 1, 2, \ldots$ he or she observes a random variable $X_t = \theta + \epsilon_t$, where $\{\epsilon_t\}$ is a sequence of independent "noise" terms, each distributed $\mathcal{N}(0, 1)$. Hereafter the random variable X_t will be referred to as a *sample*, and the process of observing successive X_t values will be called *sampling*. The strong law of large numbers says that $(X_1 + \cdots + X_t)/t \to \theta$ almost surely as $t \to \infty$, so the value of θ will eventually be revealed, but estimates of θ based on limited sampling are often important. For example, the decision maker may be considering an investment whose expected return depends on θ, and delaying the decision may have negative consequences because of discounting, or because there are direct costs associated with continued sampling. In such problems one is led to ask after each sample whether current knowledge of θ is "good enough" to justify an immediate acceptance or rejection of the investment opportunity, as opposed to continued sampling.

In the first five sections of this chapter we consider a continuous-time version of the model described above, in which the decision maker observes $Y_t = \theta t + W_t$, $t \geq 0$, where W is a standard Brownian motion. We adopt a Bayesian framework, so the decision maker's initial information is expressed in the form of a prior distribution for θ, and the dynamic inference problem is to determine the posterior distribution of θ given $\{Y_s, 0 \leq s \leq t\}$. For certain specially structured prior distributions, the dynamic inference problem can be solved explicitly, which further allows explicit solution of optimal stopping problems that arise naturally in conjunction with the inferential model.

Sections 8.6 and 8.7 consider a more general setting where the unknown drift rate evolves as a stochastic process $\{\theta_t, t \geq 0\}$. To be specific, we consider the case where $\theta_t = m(X_t)$ and $X = \{X_t, t \geq 0\}$ is a finite-state, continuous-time Markov chain. The observed process is $Y_t = \int_0^t m(X_s)ds + W_t$, and given a probability distribution for the initial state X_0, the dynamic

137

inference problem is to determine the conditional distribution of X_t given $\{Y_s, \ 0 \le s \le t\}$. The solution of the inference problem is developed in Section 8.6 without rigorous proofs, and Section 8.7 describes the application of that general theory to a problem of change-point detection.

In the early sections of this chapter, the time parameter of each stochastic process is written as a subscript. In later sections where subscripts are needed for other purposes, the time parameter is written as a functional argument for some processes, but not necessarily for all of them. For readers who are alert to this practice it should cause no confusion.

8.1 Drift-rate uncertainty in a Bayesian framework

Let W be a standard Brownian motion on a probability space $(\Omega, \mathcal{F}, P)$, and let θ be a random variable defined on that same probability space and independent of W. Also, for $t \ge 0$ let

$$(8.1) \qquad\qquad Y_t = \theta t + W_t,$$

and let $\mathbb{F} := \{\mathcal{F}_t, \ t \ge 0\}$ be the filtration of $(\Omega, \mathcal{F})$ generated by Y (see Section A.2 for the precise meaning of that term). Assuming that $E|\theta| < \infty$, we define

$$(8.2) \qquad\qquad \mu_t := E(\theta|\mathcal{F}_t), \qquad t \ge 0,$$

and

$$(8.3) \qquad\qquad Z_t := Y_t - \int_0^t \mu_s \, ds, \quad t \ge 0.$$

The precise meaning of (8.2) is explained in Section 1.6, and the following is a direct application of Theorem 1.12 (the innovation theorem).

Proposition 8.1 *Z is a standard Brownian motion with respect to $\mathbb{F}$.*

From Proposition 8.1 we have the following important conclusion: Y is an Itô process on the filtered probability space $(\Omega, \mathcal{F}, \mathbb{F}, P)$, with Itô differential

$$(8.4) \qquad\qquad dY_t = \mu_t \, dt + dZ_t.$$

8.2 Binary prior distribution

Let us consider the particular case where

$$(8.5) \qquad\qquad P(\theta = 1) = p \quad \text{and} \quad P(\theta = 0) = 1 - p,$$

with $0 < p < 1$. That is, we consider in this section the case where θ is known to lie in the set $\{0, 1\}$, so the decision maker's prior distribution is specified by a single constant p. Let us further define

$$(8.6) \qquad \xi_t := \exp\left(Y_t - \tfrac{1}{2}t\right) \quad \text{and} \quad \pi_t := h(\xi_t) \qquad \text{for } t > 0,$$

where

$$(8.7) \qquad h(x) := \frac{px}{px + (1 - p)} \qquad \text{for } 0 < x < 1.$$

The following proposition spells out explicitly the conditional distribution of θ given Y_t. It is proved by a straightforward application of Bayes' rule.

Proposition 8.2 $\pi_t = P(\theta = 1|Y_t),\ t \geq 0.$

Proof   For $t > 0$ we define the conditional density function

$$f_t^0(y) = P(Y_t \in dy|\theta = 0) = (2\pi t)^{-1/2} \exp(-y^2/2t), \qquad y \in \mathbb{R},$$

and

$$f_t^1(y) = P(Y_t \in dy|\theta = 1) = (2\pi t)^{-1/2} \exp\left(-(y - t)^2/2t\right), \qquad y \in \mathbb{R}.$$

Now Bayes' rule gives

$$(8.8) \qquad P(\theta = 1|Y_t) = \frac{pf_t^1(Y_t)}{pf_t^1(Y_t) + (1 - p)f_t^0(Y_t)}.$$

Upon substitution of the formulas for $f_t^0(Y_t)$ and $f_t^1(Y_t)$, the right side of (8.8) reduces to the formula (8.6) for π_t. □

We now show that the conditional distribution of θ given the entire history $\{Y_s,\ 0 \leq s \leq t\}$ coincides with the conditional distribution of θ given just Y_t (that is, Y_t is a *sufficient statistic* for θ). This is a direct consequence of Theorem 1.17 (the change of measure theorem), where the Radon–Nikodym derivative ξ depends on the underlying process $\{X_t,\ 0 \leq t \leq T\}$ only through its terminal value X_T.

Proposition 8.3 $\pi_t = P(\theta = 1|\mathcal{F}_t),\ t \geq 0.$

Remark   With the binary prior distribution (8.5), our general definition (8.2) of μ_t reduces to $\mu_t = P(\theta = 1|\mathcal{F}_t)$, so Proposition 8.3 gives the following in our current context:

$$(8.9) \qquad \mu_t = \pi_t, \qquad t \geq 0.$$

Proof Let $t > 0$ and $A \in \mathcal{F}_t$ be fixed but arbitrary. By the definition of conditional expectation it suffices to prove that

$$(8.10) \qquad \int_A \pi_t \, dP = \int_A 1_{\{\theta=1\}} \, dP.$$

With t fixed, we view Y as a continuous stochastic process on the restricted time domain $[0, t]$, and hence view the distribution of Y (see Section A.2) as a probability measure Q on the Borel subsets of $C[0, t]$. Specifically, (8.1) and (8.5) identify Q as the probability measure

$$(8.11) \qquad Q(\cdot) = pQ^1(\cdot) + (1 - p)Q^0(\cdot),$$

where $Q^0(\cdot)$ and $Q^1(\cdot)$ are the distributions of $(0, 1)$ Brownian motion and $(1, 1)$ Brownian motion, respectively. Because $\mathbb{F}$ is the filtration generated by Y, the set $A \in \mathcal{F}_t$ has the form

$$(8.12) \qquad A = \{\omega \in \Omega : Y \in B\}$$

for some Borel subset B of $C[0, t]$, so the right side of (8.10) can be written as follows:

$$(8.13) \qquad \int_A 1_{\{\theta=1\}} \, dP = P(\theta = 1, \ Y \in B) = pQ^1(B).$$

To similarly re-express the left side of (8.10), first observe that (8.6) can be written as

$$(8.14) \qquad \pi_t := h(g(Y_t)) \qquad \text{where } g(u) := e^{u - \frac{1}{2}t}$$

and $h(\cdot)$ is defined by (8.7). Rewriting the left side (8.10) as an integral with respect to the distribution of Y, we thus have

$$(8.15) \qquad \begin{aligned} \int_A \pi_t \, dP &= \int_B h(g(Y_t)) \, dQ \\ &= \int_B h(g(Y_t)) \left(p \, dQ^1 + (1 - p) \, dQ^0 \right). \end{aligned}$$

Theorem 1.17 (the change of measure theorem) says that $dQ^1 = g(Y_t) dQ^0$, so we have

$$(8.16) \quad h(g(Y_t)) \left(p dQ^1 + (1 - p) dQ^0 \right) = h(g(Y_t)) (pg(Y_t) + 1 - p) \, dQ^0,$$

and further substituting the definition (8.7) of $h(\cdot)$ reduces (8.16) to the following:

$$(8.17) \qquad h(g(Y_t)) \left(p dQ^1 + (1 - p) dQ^0 \right) = pg(Y_t) dQ^0.$$

As noted immediately above, $g(Y_t)dQ^0 = dQ^1$ by the change of measure theorem, so (8.15) reduces to

$$(8.18) \qquad \int_A \pi_t \, dP = \int_B p \, dQ^1 = pQ^1(B).$$

Combining (8.13) and (8.18) proves (8.10), as required. □

Proposition 8.4 $\pi = \{\pi_t, \ t \geq 0\}$ *satisfies the stochastic differential equation*

$$(8.19) \qquad d\pi = \pi(1 - \pi)dZ,$$

where Z is the standard Brownian motion defined by (8.3).

Remark It follows from (8.19) that π is a martingale with respect to the filtration $\mathbb{F}$ that is generated by Y, but that conclusion is just a particular case of the following general fact: if X is an integrable random variable on some probability space $(\Omega, \mathcal{F}, P)$, and if $\mathbb{F} = \{\mathcal{F}_t, \ t \geq 0\}$ is *any* filtration of $(\Omega, \mathcal{F})$, then the process $M_t := E(X|\mathcal{F}_t)$ is by definition a martingale with respect to $\mathbb{F}$.

Proof First observe that the function h defined in (8.7) satisfies

$$(8.20) \quad h'(x) = h(x)(1 - h(x))/x \quad \text{and} \quad h''(x) = -2h^2(x)(1 - h(x))/x^2.$$

Also, given the Itô representation (8.4) for Y, we can apply Itô's formula and the definition (8.6) of ξ to conclude that

$$(8.21) \qquad d\xi = \xi\left(dY - \tfrac{1}{2}dt\right) + \tfrac{1}{2}\xi\left(dY - \tfrac{1}{2}dt\right)^2 = \xi dY,$$

and hence that $(d\xi)^2 = \xi^2(dY)^2 = \xi^2 dt$. Thus, using the definition $\pi_t := h(\xi_t)$ and (8.20), another application of Itô's formula gives

$$\begin{aligned}
d\pi &= h'(\xi)d\xi + \tfrac{1}{2}h''(\xi)(d\xi)^2 \\
&= h(\xi)(1 - h(\xi))\,dY - h^2(\xi)(1 - h(\xi))\,dt \\
&= \pi(1 - \pi)(dY - \pi dt) = \pi(1 - \pi)dZ.
\end{aligned}$$

□

8.3 Brownian sequential detection

The situation modeled in the previous section can be described in the following terms. One of two competing *hypotheses* is known to be true (that is, either $\theta = 0$ or $\theta = 1$), and sequential sampling provides imperfect but increasingly sharp information about which one pertains. We now consider a decision maker who must choose, based on observation of the process

Y, a time T at which to stop sampling, and must decide at that time which hypothesis to "accept." If the accepted hypothesis is in fact true, then no penalty is incurred, but if it is false the penalty is as follows:

$$(8.22) \qquad \text{penalty} = \begin{cases} c_0 & \text{if hypothesis 0 is accepted but } \theta = 1 \\ c_1 & \text{if hypothesis 1 is accepted but } \theta = 0 \end{cases},$$

where $c_0, c_1 > 0$. In addition, the decision maker incurs cost at rate $c > 0$ while sampling continues.

To express this decision problem in precise mathematical terms, we first specify that T must be a stopping time with regard to the filtration $\mathbb{F}$ that is generated by Y, and further require that $E(T) < \infty$. (A stopping time with infinite expected value will incur infinite expected sampling cost and is therefore uninteresting.) For reasons explained in the next paragraph, we express the decision maker's objective as follows: choose T to

$$(8.23) \qquad\qquad \text{minimize } E\left[g(\pi_T) + cT\right],$$

where

$$(8.24) \qquad g(x) := c_0 x \wedge c_1(1 - x) \qquad \text{for } 0 < x < 1.$$

To understand the rationale for this formulation, let the stopping time T be fixed, let $x \in (0, 1)$ be arbitrary, and suppose that the posterior probability of the event $\{\theta = 1\}$ at the time of stopping is $\pi_T = x$. Then the decision maker's conditional expected penalty, given all information available at the time of stopping, is $c_0 x$ if hypothesis 0 is accepted, and is $c_1(1 - x)$ if hypothesis 1 is accepted. The decision maker will accept whichever hypothesis has the smaller conditional expected penalty, and so we arrive at the "stopping cost" $g(\cdot)$ specified in (8.24).

The optimal stopping problem (8.23) is usually described in the literature of statistics as one of "sequential hypothesis testing", but the term "sequential detection" is more common in engineering. The latter terminology reflects a conceptualization in which a decision maker (possibly a programmed device) strives to "detect" the underlying state θ.

The stopping problem (8.23) does *not* fit within the general formulation laid out in Section 5.1, because it involves minimization of positive costs rather than maximization of positive rewards (this is admittedly a minor distinction), and because the process π that describes the state of the system is not an ordinary Brownian motion (this is a more substantial distinction). Nonetheless, the basic reasoning developed in Chapter 5 applies with little change to the problem at hand, so we shall be brief in its analysis. To begin, we reduce (8.23) to an equivalent problem that has a different "stopping cost" $\hat{g}(\cdot)$ and no "sampling cost," exactly as in Section 5.2. For that purpose let

$$(8.25) \qquad f(x) := 2(2x - 1)\log\left(\frac{x}{1 - x}\right), \qquad 0 < x < 1,$$

and

$$(8.26) \qquad \hat{g}(x) := g(x) + cf(x), \qquad 0 < x < 1.$$

Proposition 8.5 *If T is a stopping time with $E(T) < \infty$, then we have*

$$(8.27) \qquad E\left[f(\pi_T)\right] = E(T) + f(p),$$

and hence

$$(8.28) \qquad E\left[g(\pi_T) + cT\right] = E\left[\hat{g}(\pi_T)\right] - cf(p).$$

Remark Thus T achieves the minimum in (8.23) if and only if it minimizes $E[\hat{g}(\pi_T)]$, and we shall adopt the latter, reduced objective hereafter.

Proof The two key properties of f that we shall use (both easy to verify) are the following:

$$(8.29) \qquad f''(x) = 2\left[x(1 - x)\right]^{-2}, \qquad 0 < x < 1,$$

and

$$(8.30) \qquad x(1 - x)f'(x) \text{ is bounded on } (0, 1).$$

Combining Propositions 8.1 and 8.4 with Itô's formula (Section 4.6), and further using (8.29), we have that

$$(8.31) \qquad \begin{aligned} df(\pi) &= f'(\pi)\pi(1 - \pi)dZ + \tfrac{1}{2}f''(\pi)\pi^2(1 - \pi)^2 dt \\ &= f'(\pi)\pi(1 - \pi)dZ + dt. \end{aligned}$$

Let T be a stopping time with $E(T) < \infty$. Integrating both sides of (8.31) between 0 and T, and then taking expectations, we have from (8.30) and Corollary 4.8 (the zero expectation property of the stochastic integral) that $E[f(\pi_T)] - f(p) = E(T)$, as required. □

To continue the analysis it will be useful to define

$$(8.32) \qquad g_0(x) = c_0 x + cf(x) \quad \text{and} \quad g_1(x) = c_1(1 - x) + cf(x)$$

for $0 < x < 1$. From (8.25) and (8.29) we see that the function $f(\cdot)$ is strictly convex on $(0, 1)$ and symmetric about $x = \tfrac{1}{2}$. Thus $g_0(\cdot)$ is strictly convex and reaches its minimum value at a point $x \in (0, \frac{1}{2})$ where $f'(x) = -c_0/c$. Similarly, $g_1(\cdot)$ is strictly convex and reaches its minimum at a point $x \in (\frac{1}{2}, 1)$ where $f'(x) = c_1/c$. It is immediate from the definitions (8.24) and (8.26) that

$$(8.33) \qquad \hat{g}(x) = g_0(x) \wedge g_1(x), \qquad 0 < x < 1.$$

Now let V be the lower convex envelope of $\hat{g}$ on $(0, 1)$, that is, the pointwise supremum of convex functions v on $(0,1)$ such that $v(x) \leq g_0(x) \wedge g_1(x)$, $0 < x < 1$. The proof of the following is left as an exercise, using (8.33) and properties of g_0 and g_1 stated immediately above. The lower convex envelope $V(\cdot)$ is pictured in Figure 8.1 as a darkly shaded curve.

Proposition 8.6   *The lower convex envelope V is C^1 and piecewise C^2 on $(0, 1)$, and has the following specific structure: there exist points x_0 and x_1 satisfying $0 < x_0 < x_1 < 1$ such that*

$$(8.34) \qquad V(x) = \begin{cases} g_0(x) & \text{if } 0 < x \leq x_0 \\ \left(\frac{x_1-x}{x_1-x_0}\right)g_0(x_0) + \left(\frac{x-x_0}{x_1-x_0}\right)g_1(x_1) & \text{if } x_0 \leq x \leq x_1 \\ g_1(x) & \text{if } x_1 \leq x < 1. \end{cases}$$

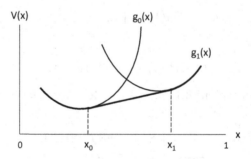

Figure 8.1 The lower convex envelope $V(\cdot)$.

The two parts of the following proposition are analogous to general results cited in Section 5.5 on the optimal stopping of an ordinary Brownian motion. The first part says that the lower convex envelope $V(\cdot)$ provides a lower bound on the expected cost achievable with any stopping time, and the second part says that the bound is achieved by the following specific stopping time:

$$(8.35) \qquad T^* := \inf \{t \geq 0 : \pi_t \leq x_0 \text{ or } \pi_t \geq x_1\}.$$

That is, the optimal continuation region is the open interval (x_0, x_1), over which the convex envelope V is linear and lies strictly below $\hat{g}$; the optimal stopping region is the complementary set where V and $\hat{g}$ coincide.

Proposition 8.7 $E[\hat{g}(\pi_T)] \geq V(p)$ *for any stopping time T with $E(T) <$* ∞, *and* $E[\hat{g}(\pi_{T^*})] = V(p)$. *Thus T^* is optimal.*

Remark   Because $c_0 x < c_1(1 - x)$ for all $x \in (0, x_0)$, the optimal choice is to accept hypothesis 0 if $\pi_{T^*} \in (0, x_0]$ and to accept hypothesis 1 if $\pi_{T^*} \in [x_1, 0)$.

Proof   First, it follows easily from (8.30), (8.32), and the structure (8.34) of V that

(8.36)
$$V''(x) = \begin{cases} cf''(x) & \text{if } 0 < x < x_0 \\ 0 & \text{if } x_0 < x < x_1 \\ cf''(x) & \text{if } x_1 < x < 1 \end{cases}$$

and

(8.37) $x(1 - x)V'(x)$ is bounded on $(0, 1)$.

Thus, proceeding exactly as in the proof of Proposition 8.5, we have that

(8.38)
$$dV(\pi) = V'(\pi)\pi(1 - \pi)dZ + \tfrac{1}{2}V''(\pi)\pi^2(1 - \pi)^2 dt$$
$$\geq V'(\pi)\pi(1 - \pi)dZ.$$

(Because V is C^1 and piecewise C^2, the standard Itô formula still applies; see Proposition 4.16.) Integrating both sides of (8.38) from 0 to T and taking expectations, we have from (8.37) and Corollary 4.8 that $E[V(\pi_T)] - V(p) \geq 0$ for any stopping time T with $E(T) < \infty$. But $\hat{g}(\cdot) \geq V(\cdot)$ as a matter of definition, so $E[\hat{g}(\pi_T)] \geq V(p)$. This proves the first statement.

Now consider the particular case $T = T^*$. Because $V''(\cdot) = 0$ in (x_0, x_1), the argument above specializes to give $E[V(\pi_{T^*})] = V(p)$. But $V(\pi_{T^*}) = \hat{g}(\pi_{T^*})$ almost surely, so the second statement is proved as well.   □

8.4 General finite prior distribution

Generalizing the model of Section 8.2, we now suppose that the observed process Y has the form

(8.39) $Y_t = m(X)t + W_t, \qquad t \geq 0,$

where W is a standard Brownian motion, X is a random variable independent of W that takes values $1, \ldots, n$ with probabilities $p_1, \ldots, p_n$, and

$m(1), \ldots, m(n)$ are given constants. This scenario fits within the general framework described in Section 8.1, but we have chosen to represent the unknown drift rate θ as $\theta = m(X)$; this overly elaborate notation foreshadows a still more general model to be introduced in Section 8.6. In the development to follow, the potential drift rates $m(1), \ldots, m(n)$ will alternatively be written as $m_1, \ldots, m_n$ when that is more convenient.

In the current context we define $\pi_i(t) := P(X = i|\mathcal{F}_t)$, $t \geq 0$, where $\mathbb{F} = \{\mathcal{F}_t, t \geq 0\}$ is the filtration generated by the observed process Y, as before. These conditional probabilities are given by explicit formulas that generalize (8.6) and (8.7), namely,

$$(8.40) \qquad \pi_i(t) = \frac{p_i \xi_i(t)}{\sum_{j=1}^n p_j \xi_j(t)} \qquad \text{for } i = 1, \ldots, n,$$

where

$$(8.41) \qquad \xi_i(t) = \exp\left(m_i Y_t - \tfrac{1}{2}m_i^2 t\right).$$

To establish these explicit formulas one can argue exactly as in Section 8.2, first defining the density functions

$$f_t^i(y) = (2\pi t)^{-1/2} \exp\left(-(y - m_i t)^2/2t\right), \qquad y \in \mathbb{R},$$

then observing that

$$(8.42) \qquad P(\theta = i|Y_t) = \frac{p_i f_t^i(y_t)}{\sum_{j=1}^n p_j f_t^j(Y_t)}.$$

The right side of (8.42) reduces to the right side of (8.40), and finally, it follows from Theorem 1.17 (the change of measure theorem) that Y_t is a sufficient statistic for $\{Y_s, \ 0 \leq s \leq t\}$ so the left side of (8.42) equals the conditional expectation that defines $\pi_i(t)$.

We can now generalize Proposition 8.4 to derive a stochastic differential equation for any one of the processes π_i. Specifically focusing on π_1 for illustration, let

$$(8.43) \qquad h(x) := \frac{p_1 x_1}{\sum_{j=1}^n p_j x_j}$$

for $x = (x_1, \ldots, x_n)$ such that $0 < x_j < 1$ for all j. Writing $\xi(t) = (\xi_1(t), \ldots, \xi_n(t))$, we then have $\pi_1(t) = h(\xi(t))$ by (8.40). Itô's formula then gives

$$(8.44) \qquad d\pi_1 = \sum_{i=1}^n h_i(\xi) \, d\xi_i + \frac{1}{2} \sum_{i=1}^n \sum_{j=1}^n h_{ij}(\xi) \, d\xi_i d\xi_j,$$

where $h_i(\cdot)$ and $h_{ij}(\cdot)$ denote first-order and second-order partial derivatives of $g(\cdot)$ as usual. Also, generalizing (8.21), we have from (8.41) and Itô's formula that

$$(8.45) \qquad d\xi_i = m_i \xi_i \, dY \qquad \text{for } i = 1, \ldots, n,$$

and hence

$$(8.46) \qquad d\xi_i d\xi_j = m_i m_j \xi_i \xi_j \, dt \qquad \text{for } i, j = 1, \ldots, n.$$

After computing the partial derivatives of h, substituting (8.45) and (8.46) and simplifying (this is a long and tedious process, with many opportunities for error), one arrives at the following re-expression of (8.44):

$$(8.47) \qquad d\pi_1(t) = \left(m_1 - \sum_{i=1}^{n} m_i \pi_i(t) \right) \pi_1(t) \left(dY - \sum_{j=1}^{n} m_j \pi_j(t) \, dt \right).$$

(Here the time parameter for stochastic processes, which was suppressed in (8.44), is restored to emphasize the distinction between constants and processes.) Finally, observe that the general definition (8.2) specializes in our current setting to

$$(8.48) \qquad \mu_t = \sum_{i=1}^{n} m_i \pi_i(t), \qquad t \geq 0,$$

so the final factor on the right side of (8.47) is the differential of the process Z defined in (8.3), which is a standard Brownian motion with respect to $\mathbb{F}$ by Proposition 8.1. Of course, a similar differential relationship holds for $\pi_2(t), \ldots, \pi_n(t)$, so we arrive at the following.

Proposition 8.8 π_i *satisfies the stochastic differential equation*

$$(8.49) \qquad d\pi_i(t) = (m_i - \mu_t) \pi_i(t) \, dZ_t \qquad \text{for } i = 1, \ldots, n,$$

where μ_t is defined by (8.48) and Z is a standard Brownian motion with respect to the filtration $\mathbb{F}$ that is generated by the observed process Y.

As indicated earlier, Proposition 8.8 generalizes Proposition 8.4 (which is specific to the case $n = 2$) and is itself subsumed by a still more general result to be developed in Section 8.6. In that more general setting, explicit formulas for the posterior probabilities $\pi_i(t)$ are not available, but (8.49) extends in a simple way, and (8.49) is all that one really needs for purposes of application.

8.5 Gaussian prior distribution

Let us consider again the general setting of Section 8.1, where the drift rate θ is fixed but uncertain. The other tractable assumption that is commonly used in applications, apart from the binary prior distribution discussed in Section 8.2, is that of a normal or Gaussian prior. The main result for that case, stated precisely in Proposition 8.9 below, is that the posterior distribution of θ at each time $t > 0$ is also Gaussian, with a mean μ_t that depends on the observed path of Y only through its terminal value Y_t, and with a variance σ_t^2 that is a deterministic function of t. The proof will only be sketched, as it is similar in structure to that for the case of a binary prior.

Proposition 8.9 *If θ is distributed $N(\mu_0, \sigma_0^2)$, where $\sigma_0 > 0$, then for each $t > 0$ one has*

$$(8.50) \qquad P(\theta \le u | \mathcal{F}_t) = \Phi\left(\frac{u - \mu_t}{\sigma_t}\right), \qquad u \in \mathbb{R},$$

where $\Phi(\cdot)$ is the $N(0, 1)$ distribution function,

$$(8.51) \qquad \mu_t = \left(\mu_0 \sigma_0^{-2} + Y_t\right)\sigma_t^2 \quad and \quad \sigma_t^2 = \left(\sigma_0^{-2} + t\right)^{-1}.$$

Remark   Our current notation is consistent with the earlier definition (8.2).

The key for proving Proposition 8.9 is the following fact from elementary probability theory: if X_0 and X_1 are independent random variables distributed $N(\mu_0, \sigma_0^2)$ and $N(0, \sigma_1^2)$, respectively, then the conditional distribution of X_0 given $S := X_0 + X_1$ is $N(\hat{\mu}, \hat{\sigma}^2)$, where $\hat{\mu} = (\mu_0 \sigma_0^{-2} + S \sigma_1^{-2})\hat{\sigma}^2$ and $\hat{\sigma}^2 = (\sigma_0^{-2} + \sigma_1^{-2})^{-1}$. In our particular setting, given the representation (8.1) of Y_t, we make the associations $X_0 = \theta$, $X_1 = t^{-1}W_t$, and $S = t^{-1}Y_t$. This leads to the formulas (8.50) and (8.51), except that the conditioning is on Y_t rather than $\mathcal{F}_t$. To complete the proof we need that Y_t serves as a sufficient statistic for $\{Y_s, 0 \le s \le t\}$, which follows from Theorem 1.17 (the change of measure theorem) as in the proof of Proposition 8.3.

Proposition 8.9 goes back at least to Chernoff (1968). The property of normal distributions that was cited in the sketch of its proof is commonly associated with the topic of *conjugate prior distributions*; see De-Groot (2004, Sect. 9.5). Theorems 11.1 and 12.1 of Liptser and Shiryayev (2000b) together provide a much more general version of Proposition 8.9.

8.6 A partially observed Markov chain

Generalizing the model of Section 8.2, we now suppose that the observed process Y has the form

$$(8.52) \qquad Y_t = \int_0^t m(X_s)\, ds + W_t, \qquad t \geq 0,$$

where $X = \{X_t,\ t \geq 0\}$ is a continuous-time Markov chain with state space $\{1,\ldots,n\}$, W is a standard Brownian motion independent of X, and $m(1),\ldots,m(n)$ are given constants. As in Section 8.4, the drift rates $m(1)$, $\ldots, m(n)$ will alternatively be written as $m_1,\ldots,m_n$ when that is more convenient.

For $i \neq j$ we denote by $q_{ij} \geq 0$ the transition intensity from state i to state j for the Markov chain X, and then for $i = 1,\ldots,n$ we define

$$q_i := \sum_{j \neq i} q_{ij} \quad \text{and} \quad q_{ii} := -q_i.$$

This means the following. If $q_i > 0$, then the duration of each visit to state i is exponentially distributed with mean $1/q_i$, and the probability that such a visit ends with a transition to state $j \neq i$ is q_{ij}/q_i, independent of all previous history. On the other hand, if $q_i = 0$, then state i is an absorbing state for the Markov chain X. The $n \times n$ matrix $Q := (q_{ij})$, which is called the *generator matrix* or *infinitesimal generator* of the Markov chain X, is assumed to be given. Also given are initial state probabilities $p_i := P(X_0 = i)$ for $i = 1,\ldots,n$. It is assumed that $0 < p_i < 1$ for all i.

As in Section 8.2, let $\mathbb{F} = \{\mathcal{F}_t,\ t \geq 0\}$ be the filtration generated by the observed process Y. In the current context we define

$$(8.53) \qquad \pi_i(t) := P(X_t = i | \mathcal{F}_t) \qquad \text{for } i = 1,\ldots,n \quad \text{and} \quad t \geq 0,$$

and

$$(8.54) \qquad \pi(t) = (\pi_1(t),\ldots,\pi_n(t)) \qquad \text{for } t \geq 0.$$

In this section we shall develop a system of linked stochastic differential equations that describe the dynamic evolution of the time-dependent distributions $\{\pi(t),\ t \geq 0\}$. In the current context, the appropriate analogs of (8.2) and (8.3) are

$$(8.55) \qquad \mu_t := E\left(m(X_t)|\mathcal{F}_t\right) = \sum_{i=1}^n m_i \pi_i(t), \qquad t \geq 0,$$

and

(8.56) $$Z_t = Y_t - \int_0^t \mu_s \, ds, \qquad t \geq 0.$$

Again readers are referred to Section 1.6 for the precise meaning of the processes π_i in (8.53), and hence of the conditional expectation process μ in (8.55). The following is another direct application of Theorem 1.12 (the innovation theorem).

Proposition 8.10 *The process Z defined by (8.56) and (8.55) is a standard Brownian motion with respect to the filtration $\mathbb{F}$ that is generated by Y.*

The following result, originally due to Wonham (1964), is a special case of Theorem 9.1 in Liptser and Shiryayev (2000a). After Wonham's theorem has been stated, its content will be explained in intuitive terms, followed by a brief description of its practical value. A simple but illuminating application is considered in Section 8.7, which concludes this chapter.

Proposition 8.11 (Wonham's theorem) *The conditional probabilities $\pi_j(t)$ satisfy the stochastic differential equations*

(8.57) $$d\pi_j(t) = \sum_{i=1}^n \pi_i(t) q_{ij} \, dt + \left(m_j - \mu_t \right) \pi_j(t) \, dZ_t$$

for $j = 1, \ldots, n$, where μ_t and Z are defined by (8.55) and (8.56), respectively.

Remark The component processes π_i are not in general martingales. This may appear at first glance to contradict the remark following Proposition 8.4, but it does not, because $\pi_i(t)$ is not the conditional probability of a *fixed* event given $\mathcal{F}_t$, but rather the conditional probability of an event $\{X_t = i\}$ that is itself changing with t.

For an intuitive understanding of Wonham's theorem it is useful to imagine a Bayesian observer whose personal probability assessments regarding the initial state X_0 are expressed by the vector $p = (p_1, \ldots, p_n)$, and whose revised assessments regarding the state at any later time t, given observations of the process Y, are expressed by the vector $\pi(t) = (\pi_1(t), \ldots, \pi_n(t))$. In the absence of such observations, we know from elementary Markov process theory that the revised assessments would satisfy

(8.58) $$d\pi_j(t) = \sum_{i=1}^n \pi_i(t) q_{ij} \, dt \qquad \text{for } j = 1, \ldots, n.$$

That is, over any short time interval $[t, t+\Delta t]$ the changes in the probabilities assessed for different states would be approximately proportional to Δt, namely,

$$(8.59) \qquad \Delta\pi_j(t) \simeq \sum_{i=1}^{n} \pi_i(t)q_{ij}\,\Delta t \qquad \text{for } j = 1,\ldots,n,$$

where $\Delta\pi_i(t) := \pi_i(t+\Delta t) - \pi_i(t)$. On the other hand, if the underlying state could never change (this is the case $q_{ij} \equiv 0$ considered in Section 8.4), we have the following from Proposition 8.8:

$$(8.60) \qquad \Delta\pi_j(t) \simeq \left(m_j - \mu_t\right)\pi_j(t)\,\Delta Y \qquad \text{for } j = 1,\ldots,n,$$

where $\Delta Y := Y_{t+\Delta t} - Y_t$. That is, if the underlying state is fixed but uncertain, the changes in the observer's probability assessments over a short time interval are all approximately proportional to the change in the observed process Y over that interval.

Wonham's theorem tells us that, for the general model with Markovian evolution of the underlying state, both motivations for revising the state probabilities $\pi_j(t)$ are present, and their effects are simply additive. On the one hand, the observer makes the adjustment (8.59) to account for possible Markovian state transitions, given the state probabilities $\pi(t)$ that pertained at the start of the interval; separately and additively, the observer makes the adjustment (8.60) in the initial probability vector $\pi(t)$ to account for the observed behavior of Y over the interval.

To understand the practical value of Proposition 8.11, let $f : \mathbb{R}^n \to \mathbb{R}$ be a C^2 function. We have from (8.52) and (8.56) that $(dZ)^2 = dt$, so Proposition 8.11 and the multi-dimensional Itô formula together give the following (this is another long and tedious calculation):

$$(8.61) \qquad df(\pi(t)) = dM(t) + \Gamma f(\pi(t))\,dt,$$

where

$$(8.62) \qquad M(t) := \sum_{j=1}^{n} \int_0^t f_j(\pi(s))\,\delta_j(\pi(s))\,\pi_j(s)\,dZ(s),$$

$$(8.63) \qquad \delta_j(y) := m_j - \sum_{k=1}^{n} m_k y_k \qquad \text{for } j = 1,\ldots,n \quad \text{and} \quad y \in \mathbb{R}^n,$$

and, for $y \in \mathbb{R}^n$,

$$(8.64) \qquad \Gamma f(y) := \sum_{j=1}^{n}\sum_{k=1}^{n} f_j(y)y_k q_{kj} + \frac{1}{2}\sum_{i=1}^{n}\sum_{j=1}^{n} f_{ij}(y)\delta_i(y)\delta_j(y)y_i y_j.$$

(Here f_j and f_{ij} denote partial derivatives as usual.) Combining (8.61) with Proposition 4.15 (the specialized integration by parts formula) and Corollary 4.8 (the zero expectation property of the stochastic integral) in a now familiar fashion, we arrive at the following conclusion: if $\lambda \geq 0$ and T is a stopping time with respect to $\mathbb{F}$ such that $E(T) < \infty$, then

$$(8.65) \qquad E\left[e^{-\lambda T} f\left(\pi(T)\right)\right] = f(p) + E\left[\int_0^T e^{-\lambda t}(\Gamma - \lambda) f\left(\pi(t)\right) \, dt\right].$$

To confirm (8.65) one uses, in addition to the results cited immediately above, the fact that π lives in the simplex

$$S := \left\{ y \in \mathbb{R}^n : y \geq 0 \text{ and } \sum_{k=1}^n y_k = 1 \right\},$$

with $\pi(0) = p$; thus, because f is C^2 by assumption, both f and its first-order partial derivatives are bounded on the compact set S, implying that the integrand in (8.65) is a bounded process.

In Section 4.11 it was shown that a formula precisely analogous to (8.65) allows us to calculate expected discounted cost functionals associated with a (μ, σ) Brownian motion X. Formula (8.65) plays the same role for the process π, as we shall illustrate in the following section.

8.7 Change-point detection

Consider a nonnegative random variable τ with the following distribution:

$$(8.66) \qquad\qquad P(\tau > t) = (1 - p)e^{-\alpha t}, \qquad t \geq 0,$$

where $0 < p < 1$ and $\alpha > 0$. That is, the distribution of τ has a probability mass of p concentrated on 0, and the conditional distribution of τ given $\tau > 0$ is exponential with mean $1/\alpha$. Now consider a Bayesian decision maker who observes

$$(8.67) \qquad\qquad Y_t = (t - \tau)^+ + W_t, \qquad t \geq 0,$$

where W is a standard Brownian motion independent of τ, and wishes to determine

$$(8.68) \qquad\qquad \pi(t) := P(\tau \leq t | \mathcal{F}_t), \qquad t \geq 0,$$

where $\mathbb{F} = \{\mathcal{F}_t, t \geq 0\}$ is the filtration generated by Y. For an interpretation of this problem, note that the observed process Y has drift 0 over the (possibly empty) interval $[0, \tau)$ and drift 1 thereafter. One may thus describe

τ as a *change-point*, and the observer's problem as one of *change-point detection*.

This stochastic model can be mapped into the setting of Section 8.6 by defining

$$X_t = \begin{cases} 1 & \text{for } 0 \le t < \tau, \\ 2 & \text{for } \tau \le t < \infty. \end{cases}$$

Then X is a two-state Markov chain whose generator matrix $Q = (q_{ij})$ is given by

$$Q = \begin{pmatrix} -\alpha & \alpha \\ 0 & 0 \end{pmatrix},$$

and the state-dependent drift rates for the observed process Y are $m_1 = 0$ and $m_2 = 1$. The initial state probabilities are $p_2 = p$ and $p_1 = (1 - p)$, and the process $\pi(t)$ defined by (8.68) coincides with what was called $\pi_2(t)$ in Section 8.6. Specializing Proposition 8.11 to the current model, one has

(8.69) $$d\pi = \alpha(1 - \pi)\,dt + \pi(1 - \pi)\,dZ,$$

so we associate with π the differential operator

(8.70) $$\Gamma f(x) := \alpha(1 - x)f'(x) + \tfrac{1}{2}x^2(1 - x)^2 f''(x),$$

and (8.65) gives the following when the interest rate λ is set to zero:

(8.71) $$E\left[f\left(\pi(T)\right)\right] = f(p) + E\left[\int_0^T \Gamma f\left(\pi(t)\right)\,dt\right]$$

for any C^2 function f and any stopping time T with $E(T) < \infty$.

Following a pattern that is by now familiar, we can use (8.71) to translate probabilistic questions related to the process π into differential equation problems. For example, let a and b be constants satisfying

(8.72) $$0 < a < p < b < 1,$$

and suppose we want to determine $P\{T(a) < T(b)\}$, where $T(a) := \inf\{t \ge 0 : \pi_t = a\}$ and similarly for $T(b)$. From (8.71) we have that

(8.73) $$P\{T(a) < T(b)\} = q(p),$$

where $q(\cdot)$ is a C^2 function on $[a, b]$ that satisfies

(8.74) $$\Gamma q(x) = 0, \qquad a \le x \le b, \text{ with } q(a) = 1 \text{ and } q(b) = 0.$$

The solution of (8.74) is

$$(8.75) \qquad q(x) := \frac{\int_x^b k(y)\,dy}{\int_a^b k(y)\,dy},$$

where

$$(8.76) \qquad k(x) := e^{-2\alpha\psi(x)}$$

and

$$(8.77) \qquad \psi(x) := \int \frac{dy}{y^2(1-y)} = \log\left(\frac{x}{1-x}\right) - \frac{1}{x}.$$

It can be shown that

$$(8.78) \qquad \int_a^b k(y)\,dy \uparrow \infty \qquad \text{as } a \downarrow 0,$$

so from (8.73) and (8.75) it follows that

$$(8.79) \qquad P\{T(0) < \infty\} = 0.$$

Other related problems, like determining $E[T(a) \wedge T(b)]$, can be solved in similar fashion.

The classical formulation of the *Brownian quickest detection problem* begins with the probabilistic model described in this section and requires the Bayesian decision maker to guess the value of τ based on observations of the process Y. That guess takes the form of a stopping time T, and the objective is to

$$(8.80) \qquad \text{minimize } E\left[1_{(\tau > T)} + c(T - \tau)^+\right],$$

where $c > 0$ is another given constant. That is, deviations of T from the actual value of τ are penalized in two ways: a lump-sum penalty of 1 is incurred (this is just a convenient choice of the monetary unit) if the decision maker stops before τ, and a cost of c per time unit is incurred as the penalty for "lateness" of T relative to τ. It is plausible that the optimal choice is $T = T(b) := \inf\{t \geq 0 : \pi_t \geq b\}$ for some $b \in (0, 1)$. The critical value b can be found, and the optimality of $T(b)$ among all stopping times can be proved, using the guess-and-verify approach outlined in Chapter 5, including the "principle of smooth fit." See Section 22.1 of Peskir and Shiryaev (2006) for details.

9

Further Examples

This chapter contains what is essentially a sequence of elaborate solved exercises. They serve to further develop themes initiated in earlier chapters, and are of three distinct types. Sections 9.1 and 9.2 describe two closely related diffusion processes that arise in particular modeling contexts. Sections 9.3 to 9.5 are more methodological in character, describing clever constructions that produce interesting variants of reflected Brownian motion. They also develop further the connection between Brownian motion and differential equation theory, providing probabilistic interpretations for certain novel boundary conditions. Section 9.6 treats a problem of dynamic capacity expansion that is closely related to the McDonald–Siegel investment model analyzed earlier in Chapter 5. Finally, Section 9.7 contains problems and complements that fill out or extend the arguments developed in earlier sections.

9.1 Ornstein–Uhlenbeck process

Let $\sigma > 0$ and $\lambda > 0$ be given constants, and let W be a standard Brownian motion. The *Ornstein–Uhlenbeck* process (or O–U process) Z with parameters σ and λ, and with initial state $Z_0 \in \mathbb{R}$, can be defined as follows:

$$(9.1) \qquad Z_t = e^{-\lambda t}\left(Z_0 + \sigma \int_0^t e^{\lambda s}\, dW_s\right), \qquad t \ge 0.$$

Because the integral on the right side of (9.1) involves a continuous integrator and a VF integrand, it can be interpreted in the ordinary Riemann–Stieltjes sense; see Section B.3. Using Proposition 4.15 (the specialized integration by parts formula where one factor is exponential) is is easy to verify that this process Z satisfies the stochastic differential equation (SDE)

$$(9.2) \qquad dZ_t = -\lambda Z_t dt + \sigma dW_t.$$

That is, the variance parameter of Z is σ^2, regardless of the current state, but its drift parameter pushes Z back toward the origin at a rate proportional to its current distance from the origin. The constant of proportionality λ might aptly be described as a *restorative force*. As we shall see shortly, the long-run mean value of Z_t is zero, and because Z is always being driven toward that mean value, it is frequently characterized as *mean reverting*.

The O–U process has found applications in physical science (for example, to model the motion of a spring subject to thermal fluctuations), in biology (specifically, as a model of neuronal response), and in mathematical finance to model the dynamics of interest rates. In all of those contexts it is useful to consider the following more general version of (9.2):

$$(9.3) \qquad dZ_t = -\lambda(Z_t - \theta)dt + \sigma dW_t,$$

where $\theta \in \mathbb{R}$ is a third process parameter. That is, the long-run mean value toward which Z reverts can be an arbitrary point $\theta \in \mathbb{R}$ rather than zero. An explicit solution of (9.3) is obtained by putting $Z_t - \theta$ in place of Z_t and $Z_0 - \theta$ in place of Z_0 in (9.1). For simplicity, the next paragraph treats only the special case $\theta = 0$; the modifications needed to treat general θ should be obvious.

By considering the Riemann–Stieltjes sums that approximate the integral on the right side of (9.1), one sees that the integral is itself a normally distributed random variable for any fixed value of t (see Problem 9.1), and hence that Z_t is also normally distributed. To be specific, with $\theta = 0$ the distribution of Z_t is $N(\mu_t, \sigma_t^2)$ where

$$(9.4) \qquad \mu_t = e^{-\lambda t} Z_0$$

and

$$(9.5) \qquad \sigma_t^2 = \left(\sigma e^{-\lambda t}\right)^2 \int_0^t \left(e^{\lambda s}\right)^2 ds = \left(\frac{\sigma^2}{2\lambda}\right)\left(1 - e^{-2\lambda t}\right).$$

Thus one has that $\mu_t \to 0$ as $t \to \infty$, as claimed earlier, and that $\sigma_t^2 \to \sigma^2/2\lambda$.

9.2 Probability of ruin with compounding assets

Consider an insurance company that has initial wealth $y > 0$ and whose cumulative net income is modeled by a (μ, σ) Brownian motion $X = \{X_t, t \geq 0\}$. That is, X_t represents the difference between premium payments received from policy holders over the interval $[0, t]$ and routine disbursements made over that period to pay claims by policy holders and for other operating expenses. Let us further suppose that the firm's wealth is continuously invested in a risk-free security (or a bank account) that pays interest at rate $\lambda > 0$, so its wealth at time t is given by the following Riemann–Stieltjes integral:

$$(9.6) \qquad Y_t = e^{\lambda t} y + \int_0^t e^{\lambda(t-s)} \, dX_s, \qquad t \geq 0.$$

Comparing (9.6) with (9.1), we see two differences between our wealth process Y and the classical O–U process Z. First, the integrator in the definition of Y is a (μ, σ) Brownian motion, whereas the integrator in the definition of Z is a $(0, \sigma)$ Brownian motion. Second, and more importantly, the positive process parameter λ appears with opposite signs in (9.1) and (9.6), so the SDE that characterizes Y is

$$(9.7) \qquad dY_t = \lambda Y_t dt + (\mu dt + \sigma dW_t).$$

The first term on the right side of (9.7) is a *repulsive force* that pushes Y away from the origin at a rate proportional to its current distance from the origin. Emanuel et al. (1975) called the diffusion process Y *compounding Brownian motion*. Now let

$$(9.8) \qquad T := \inf\{t \geq 0 : Y_t = 0\}.$$

In mathematical risk theory T is called the *time of ruin*. Our goal in this section is to calculate $P(T < \infty)$, referred to hereafter as the *probability of ruin*.

As the first step in that analysis we rewrite (9.6) as

$$(9.9) \qquad Y_t = e^{\lambda t}(y + \xi_t), \qquad t \geq 0,$$

where

$$(9.10) \qquad \xi_t := \int_0^t e^{-\lambda s} \, dX_s, \qquad t \geq 0.$$

The following simple proposition lists various properties of the process $\xi = \{\xi_t, \, t \geq 0\}$. Problems 9.1 and 9.2 develop these properties in a more general setting.

Proposition 9.1 *ξ is continuous and has independent, normally distributed increments. In particular, ξ_t is normally distributed for each fixed $t > 0$, with*

$$(9.11) \qquad E(\xi_t) = \frac{\mu}{\lambda}\left(1 - e^{-\lambda t}\right) \quad \text{and} \quad \text{Var}(\xi_t) = \frac{\sigma^2}{2\lambda}\left(1 - e^{-2\lambda t}\right).$$

Finally, $\xi_\infty := \lim \xi_t$ exists almost surely and is distributed $N(\mu/\lambda, \sigma^2/2\lambda)$.

Corollary 9.2 *Defining $H(u) := P(\xi_\infty \leq u)$, we have*

$$(9.12) \qquad H(u) = 1 - \Phi(b - au), \qquad u \in \mathbb{R},$$

where $a = \sqrt{2\lambda}/\sigma$, $b = a\mu/\lambda$, and $\Phi(\cdot)$ is the $N(0,1)$ distribution function.

Proof Immediate from the last sentence of Proposition 9.1 and the identity $\Phi(-x) = 1 - \Phi(x)$. □

To develop a neat formula for the probability of ruin, we first observe that (9.8) and (9.9) together imply

$$(9.13) \qquad T = \inf\{t \geq 0 : y + \xi_t = 0\}.$$

Next, (9.13) implies that

$$(9.14) \qquad \{y + \xi_\infty \leq 0\} = \{T < \infty\} \bigcap \{\xi_\infty - \xi_T \leq 0\}.$$

Defining

$$(9.15) \qquad \xi_t^* := e^{\lambda T}(\xi_{T+t} - \xi_T) \qquad \text{for all } t \geq 0 \text{ on } \{T < \infty\},$$

we can rewrite (9.14) as

$$(9.16) \qquad \{y + \xi_\infty \leq 0\} = \{T < \infty\} \bigcap \{\xi_\infty^* \leq 0\}$$

and it follows from Theorem 1.7 (the strong Markov property of Brownian motion) that

$$(9.17) \qquad P(T < \infty, \xi_\infty^* \leq 0) = P(T < \infty)P(\xi_\infty \leq 0);$$

see Problem 9.3. Recalling that $H(u) := P(\xi_\infty \leq u)$, we have from (9.16) and (9.17) that $H(-y) = P(T < \infty)H(0)$, and substituting formula (9.12) for $H(\cdot)$ in that identity brings us to the following final result:

$$(9.18) \qquad P(T < \infty) = \frac{1 - \Phi(b + ay)}{1 - \Phi(b)}.$$

The probabilistic argument used here to derive (9.18) is taken from Harrison (1977). Emanuel et al. (1975) used an alternative analytical approach to derive the same result earlier, and in describing it we shall denote by x a generic state of the diffusion process Y, reserving y to denote the fixed initial state. From (9.7) one sees that the coefficient functions of Y are (here we use the notation introduced in Section 4.5) $\mu(x) = \mu + \lambda x$ and $\sigma(x) = \sigma$, so the associated differential operator Γ is as follows:

(9.19) $$\Gamma f(x) = (\mu + \lambda x)f'(x) + \tfrac{1}{2}\sigma^2 f''(x).$$

Suppose that f is a C^2 function on $[0, \infty)$ satisfying the ordinary differential equation (ODE)

(9.20) $$\Gamma f(x) = 0, \qquad 0 < x < \infty,$$

with boundary conditions

(9.21) $$f(0) = 1 \quad \text{and} \quad f(x) \to 0 \qquad \text{as } x \to \infty.$$

It then follows from Itô's formula that $f(y) = P(T < \infty)$; see Problem 9.4 for elaboration. The unique solution of the ODE problem (9.20)–(9.21) is $f(x) = [1 - \Phi(b + ax)]/[1 - \Phi(b)]$, so we arrive at formula (9.18).

9.3 RBM with killing at the boundary

Let $Z = X + L - U$ be reflected Brownian motion (RBM) in $[0, b]$ as in Chapter 6, and let $T := \inf\{t \geq 0 : Z_t = b\}$. (This is the same notation that was used in Problems 6.1 to 6.6, to which reference will be made shortly.) Also, let u be a continuous function on $[0, b]$ and define

$$f(x) := E_x\left[\int_0^T u(Z_t)\, dt\right].$$

One may interpret $f(x)$ as the expected total payout, given that $Z_0 = X_0 = x$, from a contract that pays its owner at rate $u(Z_t)$ until the upper barrier b is hit. Using Itô's formula, we have seen in Chapter 6 (see in particular Problem 6.2) that f satisfies the differential equation

(9.22) $$\Gamma f(x) + u(x) = 0, \qquad 0 \leq x \leq b$$

(here $\Gamma f := \tfrac{1}{2}\sigma^2 f'' + \mu f'$ as in Chapter 6), subject to the boundary conditions

(9.23) $$f'(0) = 0 \quad \text{and} \quad f(b) = 0.$$

Now let ξ be an exponentially distributed random variable with mean 1, independent of X and hence also of Z, and define

(9.24) $$\tau := \inf\{t \geq 0 : \kappa L_t = \xi\},$$

where $\kappa > 0$ is a given constant. Because $L_t \uparrow \infty$ almost surely as $t \uparrow \infty$, we know that $\tau < \infty$ almost surely, and because L increases only at times t when $Z_t = 0$, we know that $Z_\tau = 0$.

Next, we redefine $f(\cdot)$ as follows:

$$(9.25) \qquad f(x) := E_x\left[\int_0^{T \wedge \tau} u(Z_t)\, dt \right], \qquad 0 \leq x \leq b,$$

which can be rewritten as

$$(9.26) \qquad f(x) := E_x\left[\int_0^\infty u(Z_t) 1_{(\tau > t)} 1_{(T > t)}\, dt \right], \qquad 0 \leq x \leq b.$$

What has changed in this redefinition is that payouts cease at time τ if $\tau < T$. We call τ a *killing time*, and describe $\{Z_t,\ 0 \leq t \leq \tau\}$ as RBM in $[0, b]$ *with exponential killing* (or memoryless killing) at the lower boundary. For reasons that will become apparent, κ is called the *killing rate* at the lower boundary.

With $f(\cdot)$ defined by (9.25), how should one modify the differential equation problem (9.22)–(9.23)? To answer that question, let $\mathbb{F} = \{\mathcal{F}_t,\ t \geq 0\}$ be the filtration generated by X as in Chapter 6, and rewrite (9.26) as

$$(9.27) \qquad \begin{aligned} f(x) &= E_x\left\{ \int_0^\infty E_x\left[u(Z_t) 1_{(\tau > t)} 1_{(T > t)} \middle| \mathcal{F}_t \right] dt \right\} \\ &= E_x\left\{ \int_0^\infty u(Z_t) 1_{(T > t)} P(\tau > t | \mathcal{F}_t)\, dt \right\}. \end{aligned}$$

The first equality in (9.27) is justified by Fubini's theorem and the tower property of conditional expectations. To justify the second equality, we use the fact that Z is adapted to $\mathbb{F}$ and T is a stopping time with respect to $\mathbb{F}$. By assumption, the exponential random variable η is independent of $\mathcal{F}_t$ for any fixed t, whereas L_t is measurable with respect to $\mathcal{F}_t$, so we have from (9.24) that

$$(9.28) \qquad P(\tau > t | \mathcal{F}_t) = P(\kappa L_t \leq \xi | \mathcal{F}_t) = e^{-\kappa L_t}.$$

Substituting (9.28) into (9.27) and rearranging gives

$$(9.29) \qquad f(x) = E_x\left[\int_0^T e^{-\kappa L_t} u(Z_t)\, dt \right], \qquad 0 \leq x \leq b,$$

and we have seen in Problem 6.6 that this function f satisfies the same differential equation (9.22) as before, and the same upper boundary condition $f(b) = 0$, but with the modified lower boundary condition

$$(9.30) \qquad f'(0) - \kappa f(0) = 0.$$

Problems 9.5 and 9.6 consider a more complex scenario where there are terminal rewards (one for the case $\tau < T$ and another for the case $T < \tau$) and all payouts are discounted.

The boundary condition (9.30) specifies a value for a linear combination of $f(0)$ and $f'(0)$. In differential equation theory this is called a *Robin* (or third-type) *boundary condition*. That terminology distinguishes it from a *Dirichlet* (or first-type) *boundary condition*, in which $f(0)$ is specified, and from a *Neumann* (or second-type) *boundary condition*, in which $f'(0)$ is specified.

A reflected diffusion process with exponential killing at the boundary occurs in the work of Daley and Green (2012) on dynamic markets with asymmetric information. In their game-theoretic model the diffusion process represents the posterior probability assigned by potential buyers to one of two possible states of the world (see Section 8.2), and the distinctive boundary behavior arises endogenously from the seller's use of a randomized strategy.

9.4 Brownian motion with an interval removed

Let a and b be real constants satisfying $0 < a < b < \infty$. In this section we shall construct a Markov process Y that takes values in $\mathbb{R}$ and has the following properties. First, Y never enters the open interval (a, b) and never "jumps over" any point outside that interval, so all jumps of Y, if there are any, must be either from a to b or else from b to a. Second, Y starts at zero and behaves like a standard Brownian motion outside the closed interval $[a, b]$. Finally, Y is a martingale.

The obvious way to construct such a process is to start with a standard Brownian motion W (recall from Section 1.1 that this includes the stipulation $W_0 = 0$) and "remove" time points t at which $W_t \in (a, b)$. Mathematically this is accomplished by the following random time scale transformation: for $t \geq 0$ let A_t be the Lebesgue measure of the set $\{s \in [0, t] : W_s \notin (a, b)\}$, then define an increasing process B via

$$(9.31) \qquad B_t := \inf\{s \geq 0 : A_s \geq t\}, \qquad t \geq 0.$$

It can be shown that the process $Y_t := W(B_t)$, $t \geq 0$, has all of the properties specified above, but it is not obvious how to solve concrete problems using that constructive definition. For example, given $\lambda > 0$ and a bounded, continuous function $u : \mathbb{R} \to \mathbb{R}$, how would one calculate

$$(9.32) \qquad E\left[\int_0^\infty e^{-\lambda t} u(Y_t)\, dt\right]?$$

The following alternative construction of Y is much easier to work with. It involves a sequence of steps, and only the first one needs to be described in detail. Starting with a standard Brownian motion W and defining $T := \inf\{t \geq 0 : W_t = a\}$, we set $Y_t = W_t$ for $0 \leq t < T$, but how should Y be defined thereafter? One idea would be to have Y jump immediately to b, but then symmetric reasoning would dictate that Y jump immediately back to a, which brings us back to the question of where to go from a. At the other extreme, we could impose an upper reflecting barrier at a, defining

$$(9.33) \qquad L_t := \sup_{0 \leq s \leq t}(W_s - a)^+, \qquad t \geq 0,$$

and $Y_t := W_t - L_t$, $t \geq 0$. (Here the letter L denotes the "pushing process" associated with an upper barrier at a, whereas in Section 9.3 and earlier it denoted the pushing process associated with a lower barrier at zero. This re-use of notation is intended to emphasize parallels between the development in this section and that in Section 9.3.) That process Y has all the properties originally stipulated *except* the martingale property, which it rather obviously lacks.

However, the construction with an upper barrier at a can be "fixed" by adding a feature similar to exponential killing at the boundary (Section 9.3). Let ξ be an exponentially distributed random variable with mean 1, independent of W, and define

$$(9.34) \qquad \tau := \inf\{t \geq 0 : \kappa L_t = \xi\},$$

where

$$(9.35) \qquad \kappa := (b - a)^{-1}.$$

We now define $Y_t := W_t - L_t$ for $0 \leq t < \tau$ and $Y_\tau = b$. This completes the construction of Y up until the time τ of its first jump, and then a similar procedure can be used to construct Y between time τ and the time of its second jump. That second process segment involves the imposition of a lower reflecting barrier at b, and it uses a second exponentially distributed random varible that is independent of both ξ and W. The second segment ends with a jump from b to a, and then the constructive process repeats in the obvious way.

Because L increases only when $W - L = a$, the jump occurring at time

τ is indeed from a to b. To confirm that Y has all the properties originally stipulated, it remains only to show that

(9.36) $\{Y_t,\ 0 \le t \le \tau\}$ is a martingale;

the martingale property of the entire process Y can then be proved by induction, using the strong Markov property of W. One way to make (9.36) apparent is to write

(9.37) $$Y_t = W_t + M(L_t), \qquad 0 \le t \le \tau,$$

where

(9.38) $$M_t = (b - a)N(\kappa t) - t, \qquad t \ge 0,$$

and N is a Poisson process with unit intensity, independent of W, whose first jump time is ξ. Because M is a martingale (with respect to its own filtration) and is independent of W, the desired conclusion (9.36) follows from (9.37), but the argument involves some technical details that we omit. (It must be established that each term on the right side of (9.37) is a martingale with respect to the filtration generated by Y, which in turn requires that each be adapted to that filtration.)

The version of Y constructed immediately above does indeed have the same distribution as the one defined by random time scale transformation, although proving that equivalence is not a simple task. Moreover, by combining the segment-by-segment construction of Y, relying as it does on a variant of exponential killing at the boundary, with the analytical theory developed in Section 9.3 (see also Problems 9.5 and 9.6), one can develop a complete analytical theory for Y. For example, remembering that $Y_0 = 0$ as a matter of definition, readers can verify that the expectation in (9.32) equals $f(0)$, where $f(\cdot)$ is the bounded C^2 solution of the differential equation

(9.39) $$\tfrac{1}{2}f''(x) - \lambda f(x) + u(x) = 0 \qquad \text{for } x \le a \text{ and } x \ge b,$$

with boundary conditions

(9.40) $$f'(a) = f'(b) = \kappa[f(b) - f(a)].$$

More generally, $f(x)$ represents the conditional expectation of the integral in (9.32) given that $Y_0 = x$.

9.5 Brownian motion with sticky reflection

Let $Z = X + L$ be (μ, σ) RBM on $[0, \infty)$, as in Section 6.6. Given a constant $\beta > 0$, we define strictly increasing processes A and τ as follows:

$$(9.41) \qquad\qquad A(t) := t + \beta L_t, \qquad t \geq 0,$$

and

$$(9.42) \qquad \tau_t := A^{-1}(t) = \inf\{s \geq 0 : A(s) = t\}, \qquad t \geq 0.$$

Following a now familiar pattern, we may write the time parameter of these processes, and of other processes as well, as either a subscript or a functional argument, depending on convenience. The focus of this section is the process

$$(9.43) \qquad\qquad Y_t := Z(\tau_t), \qquad t \geq 0.$$

For purposes of discussion let $t > 0$ be fixed, and suppose for the moment that $Z_0 = X_0 = 0$, which ensures that $L_t > 0$ almost surely, and hence that $\tau_t < t$ almost surely. The sample path of Y over the time interval $[0, t]$ is identical to the sample path of Z over the shorter interval $[0, \tau_t]$, except that the set $\{s \in [0, \tau_t] : Z_s = 0\}$, which has Lebesgue measure zero, is "fattened" or "dilated" to form the set $\{s \in [0, t] : Y_s = 0\}$, which has Lebesgue measure equal to $\beta L(\tau_t)$. Thus the degree of "fattening" or "dilation" is directly proportional to the parameter β.

Given $\lambda > 0$ and a bounded continuous function $u : [0, \infty) \to \mathbb{R}$, let

$$(9.44) \qquad \begin{aligned} f(x) &:= E_x\left[\int_0^\infty e^{-\lambda t} u(Y_t)\, dt\right] \\ &= E_x\left[\int_0^\infty e^{-\lambda t} u\left(Z(\tau_t)\right) dt\right], \qquad x \geq 0. \end{aligned}$$

We interpret $f(x)$ as the expected present value of earnings from a contract that pays its owner at rate $u(Y_t)$ for all $t \geq 0$, given that $Y_0 = Z_0 = x$, where λ is the interest rate for discounting.

Making the change of variable $s = \tau_t$, so that $t = A(s) = s + \beta L_s$, we can rewrite (9.44) as

$$(9.45) \qquad f(x) = E_x\left[\int_0^\infty e^{-\lambda(s + \beta L_s)} u(Z_s)\, (ds + \beta dL_s)\right].$$

Defining $\kappa := \lambda\beta$, and recalling that L only increases when $Z = 0$, (9.45) reduces to

$$(9.46) \qquad f(x) = E_x\left\{\int_0^\infty e^{-(\lambda s + \kappa L_s)}\left[u(Z_s)\, ds + \beta u(0)\, dL_s\right]\right\}.$$

The zero-set dilation referred to previously manifests itself in two ways in (9.46). First, there is "accelerated discounting" in every time interval where Z touches the boundary at zero, because the factor $\exp(-\kappa L_s)$ increases during every such interval. (Comparing (9.46) with (9.29), readers will see that the "accelerated discounting" effect is indistinguishable from exponential killing at the boundary.) Second, each visit of Z to the boundary at zero produces "extra earnings" equal to $\beta u(0)dL$.

What differential equation, with what boundary condition, does the function f in (9.46) satisfy? To answer that question, let $f : [0, \infty) \to \mathbb{R}$ be an arbitrary bounded C^2 function, and define $V_t := \exp(-\lambda t - \kappa L_t)f(Z_t)$. Using Proposition 4.14 (the specialized integration by parts formula) and (6.69), we have that

$$
\begin{aligned}
dV_t &= \exp(-\lambda t - \kappa L_t)\,(df(Z_t) - f(Z_t)(\lambda dt + \kappa dL_t)) \\
&= \exp(-\lambda t - \kappa L_t)\,[(\Gamma - \lambda)f(Z_t)dt \\
&\quad + (f'(0) - \kappa f(0))\,dL_t + \sigma f'(Z_t)dW_t],
\end{aligned}
\tag{9.47}
$$

where W is the standard Brownian motion in the representation $X_t = \mu t + \sigma W_t$ and $\Gamma f := \mu f' + \frac{1}{2}\sigma^2 f''$ as usual. Integrating both sides of (9.47) over $[0, T(b)]$, where $T(b) := \inf\{t \geq 0 : Z_t = b\}$, then taking E_x of both sides and letting $b \to \infty$, we arrive at the following:

$$
f(x) = E_x\left\{ \int_0^\infty e^{-(\lambda t + \kappa L_t)}\,[(\lambda - \Gamma)f(Z_t)\,dt \right.
$$
$$
\left. + (\kappa f(0) - f'(0))\,dL_t] \right\}, \qquad x \geq 0.
\tag{9.48}
$$

Comparing (9.46) and (9.48), we see that the desired function f must satisfy

$$
\Gamma f(x) - \lambda f(x) + u(x) = 0, \qquad x \geq 0,
\tag{9.49}
$$

with boundary condition

$$
f'(0) - \kappa f(0) = u(0).
\tag{9.50}
$$

This is another instance of a Robin (or third-type) boundary condition, specifying the value for a linear combination of $f(0)$ and $f'(0)$.

The boundary behavior of the process Y is often called "sticky reflection," because unlike the "instantaneously" reflected process Z studied in Chapter 6, Y actually occupies the boundary for a positive amount of time (that is, the set of time points t at which $Y_t = 0$ has positive Lebesgue measure) during each interval in which Y visits the boundary. Harrison and Lemoine (1981), calling the process Y "sticky Brownian motion," described its potential role as a storage system model.

9.6 A capacity expansion model

As in Section 5.3 (devoted to the McDonald–Siegel investment model), let us consider a geometric Brownian motion $Y_t = \exp(X_t)$, $t \geq 0$, where X is a Brownian motion with starting state x, variance parameter $\sigma^2 > 0$, and drift parameter $\gamma - \sigma^2/2$. Thus $E_x(Y_t) = \exp(x + \gamma t)$ for any $t \geq 0$, and $dY = \sigma Y dW + \gamma Y dt$, where W is a standard Brownian motion. We interpret Y_t as the price of some commodity (like copper) at time t.

Let us denote by I_t the cumulative amount of money that is invested over the time interval $[0, t]$ in a plant or other physical asset (like a mine) that produces the commodity: the plant's production rate at time t, in units like tons per day, is $h(I_t)$, where h is a differentiable, strictly concave function on $[0, \infty)$ with $h(0) = 0$. Thus the plant's revenue rate at time t, in units like dollars per day, is $h(I_t)Y_t$. A decision maker must choose the investment process $I = \{I_t, \ t \geq 0\}$ as a non-anticipating functional of (that is, as a process adapted to the filtration generated by) the commodity price process Y_t, or equivalently, as a non-anticipating functional of the underlying Brownian motion X. Given an interest rate $\lambda > \gamma$ for discounting, the objective is to maximize the expected present value of revenues received minus investment dollars spent over an infinite planning horizon.

This problem appears at first blush to be more complicated than the single-shot McDonald–Siegel problem, but it actually is not. As an aid to intuition, consider an alternative scenario where the function $h(\cdot)$, instead of being continuous, is piecewise constant with $h(0) = 0$ and with jumps of size $\delta_1, \delta_2, \ldots$ at cumulative investment levels of $\Delta, 2\Delta, \ldots$, where $\Delta > 0$ and $\delta_1 > \delta_2 > \cdots > 0$. That is, in the alternative scenario, investment must come in discrete "chunks" of size Δ, and successive chunks of investment provide successively smaller increases in the production rate. If we consider any one chunk in isolation, ignoring the constraint that the nth chunk of investment can only occur after the first $n - 1$ have occurred, then the analysis in Section 5.3 can be applied, as follows.

If the nth chunk of investment is made at a time t when $X_t = b$, then as shown at the beginning of Section 5.3, the future revenues that it provides have expected present value $\delta_n k \exp(b)$, where $k := (\lambda - \gamma)^{-1} > 0$. That gain must be balanced against the immediate investment of Δ dollars, so Proposition 5.4 tells us that the optimal time to make the investment is

$T(b_n)$, where

(9.51) $$T(z) := \inf\{t \geq 0; \, X_t \geq z\}, \qquad z \in \mathbb{R},$$

b_n satisfies

(9.52) $$\delta_n k \exp(b_n) = \left(\frac{\alpha_2}{\alpha_2 - 1}\right)\Delta,$$

and $\alpha_2 := \alpha_2(\lambda)$ is given by formula (3.23). Because $\delta_1 > \delta_2 > \ldots$, the X values b_n that trigger successive investment increments automatically satisfy $0 \leq b_1 \leq b_2 \leq \ldots$, and hence $0 \leq T(b_1) \leq T(b_2) \leq \ldots$. Thus the constraint on the timing of investment chunks is satisfied automatically, and we conclude that the trigger values b_n defined by (9.52) constitute an optimal strategy for the alternative scenario with discrete investment chunks.

To approximate in this way the original problem where $h(\cdot)$ is continuous, let $\Delta > 0$ be small and define $\delta_n := h(n\Delta) - h((n-1)\Delta) \simeq h'(n\Delta)\Delta$. Thus for small Δ the trigger value b_n that brings cumulative investment to level $n\Delta$ satisfies

(9.53) $$h'(n\Delta)k \exp(b_n) \simeq \left(\frac{\alpha_2}{\alpha_2 - 1}\right).$$

Letting $\Delta \downarrow 0$, we are led to conjecture that the following investment strategy is optimal for our original problem: define $b(\cdot)$ via

(9.54) $$h'(u)ke^{b(u)} = \left(\frac{\alpha_2}{\alpha_2 - 1}\right) \qquad \text{for } u \geq 0,$$

and then let I be the continuous increasing process defined by the relationship $\{I_t \leq u\} = \{T(b(u)) \geq t\}$ for $u \geq 0$. Defining $M_t := \sup\{X_s, \, 0 \leq s \leq t\}$ as in earlier chapters, we can restate this as

(9.55) $$I_t = b(M_t), \qquad t \geq 0.$$

To derive the putative optimal strategy (9.55), we started by making a piecewise constant *lower* approximation to the continuous, concave function $h(\cdot)$. The best achievable expected present value in that discrete approximation is a lower (pessimistic) bound on the best achievable value in our original problem formulations. In similar fashion, one can make a piecewise constant *upper* approximation to $h(\cdot)$, solve the resulting investment problem exactly, and thereby derive an upper (optimistic) bound on the true optimal value. By comparing the optimal solutions and optimal values in the two approximations, then letting the discretization parameter Δ approach zero in each, one obtains a rigorous proof that (9.55) is indeed an optimal investment policy strategy in our original problem formulation.

9.7 Problems and complements

Problem 9.1   Let $G : [0, \infty) \to \mathbb{R}$ be a continuous VF function and define

$$(9.56) \qquad\qquad \xi_t := \int_0^t G_s \, dW_s, \qquad t \geq 0,$$

where W is a standard Brownian motion. Show that ξ has independent, normally distributed increments, and in particular that $\xi_t \sim \mathcal{N}(0, \sigma_t^2)$, where

$$\sigma_t^2 = \int_0^t G_s^2 \, ds, \qquad t \geq 0.$$

Hint: Consider a conveniently chosen sequence of Riemann–Stieltjes sums that approximate the integral in (9.56).

Problem 9.2 (*Continuation*)   Assuming $\sigma_\infty^2 < \infty$, use the martingale convergence theorem (see Section A.4) to show that $\xi_\infty := \lim \xi_t$ exists almost surely and is distributed $\mathcal{N}(0, \sigma_\infty^2)$.

Problem 9.3   In the setting of Section 9.2, where T is defined by (9.8), let $a > 0$ be fixed and define X^* as in Theorem 1.7 on $\{T < a\}$, observing that

$$(9.57) \qquad\qquad \xi_t^* = \int_0^t e^{-\lambda s} \, dX_s^* \qquad \text{for } t \geq 0 \text{ on } \{T < a\}.$$

Use Theorem 1.7 (the strong Markov property of Brownian motion) to show that

$$P(T < a, \ \xi_\infty^* \leq 0) = P(T < a)P(\xi_\infty \leq 0),$$

and then let $a \uparrow \infty$ to establish (9.17).

Problem 9.4   With Y and T defined as in Section 9.2, let $b > y$ be fixed and define $T(b) := \inf\{t \geq 0 : Y_t = b\}$. From (9.9) and the result of Problem 9.2, it follows both that $|Y_t| \to \infty$ almost surely as $t \to \infty$, implying that $T \wedge T(b) < \infty$ almost surely, and that $T(b) \to \infty$ almost surely as $b \to \infty$. Now let f be as described immediately after (9.19). Use the differential characterization (9.7) of Y, plus the result of Problem 4.18, to show that

$$f(y) = P\{T < T(b)\} + f(b)P\{T > T(b)\}.$$

Now let $b \uparrow \infty$ to conclude that $f(y) = P(T < \infty)$.

Problem 9.5   In the setting of Section 9.3, let $\lambda \geq 0$ be fixed and let f be a C^2 function on $[0, b]$. Use the expression (9.47) for the Itô differential of the process $V_t := \exp(-\lambda t - \kappa Z_t) f(Z_t)$ to conclude the following:

$$(9.58) \quad E_x\left[e^{-\lambda T}e^{-\kappa L_T}f(b)\right] = f(x) - E_x\left[\int_0^T e^{-\lambda t}e^{-\kappa L_t}u(Z_t)\,dt\right]$$

$$- [\kappa f(0) - f'(0)]\, E_x\left(\int_0^T e^{-\lambda t}e^{-\kappa L_t}\,dL_t\right), \qquad 0 \le x \le b,$$

where $u(z) := \lambda f(z) - \Gamma f(z)$, $0 \le z \le b$.

Problem 9.6 (*Continuation*) Given a continuous function $u : [0,b] \to \mathbb{R}$, real constants α and β, and an interest rate $\lambda \ge 0$ for discounting, suppose we want to calculate

$$(9.59) \quad f(x) := E_x\left[\int_0^{T \wedge \tau} e^{-\lambda t}u(Z_t)\,dt\right]$$

$$+ \alpha E_x\left(e^{-\lambda\tau};\ \tau < T\right) + \beta E_x\left(e^{-\lambda T};\ T < \tau\right), \qquad 0 \le x \le b.$$

One may interpret $f(x)$ as the expected present value, given that $Z_0 = X_0 = x$, of a contract which pays its owner at rate $u(Z_t)$ until either level b is hit or else killing occurs, pays α at the time of killing if killing occurs before level b is hit, and pays β at the first hitting time of level b if killing has not occurred before then. Show that the three terms on the right side of (9.59) are equal to the following expressions:

$$(9.60) \qquad \text{1st term} = E_x\left[\int_0^T e^{-\lambda t}e^{-\kappa L_t}u(Z_t)\,dt\right],$$

$$(9.61) \qquad \text{2nd term} = \alpha E_x\left(\int_0^T e^{-\lambda t}\kappa e^{-\kappa L_t}\,dL_t\right),$$

$$(9.62) \qquad \text{3rd term} = \beta E_x\left(e^{-\lambda T}e^{-\kappa L_T}\right).$$

Combined with the result in Problem 9.5, this shows the following: to compute the expectation $f(\cdot)$ defined by (9.59), it suffices to solve the differential equation $\lambda f(x) - \Gamma f(x) = u(x)$, $0 \le x \le b$, subject to boundary conditions $f'(0) = \kappa[f(0) - \alpha]$ and $f(b) = \beta$.

Appendix A

Stochastic Processes

The first three sections of this appendix are concerned with notation and terminology. Readers should particularly note the standing assumptions such as joint measurability of stochastic processes. The last two sections are brief, stating without proof a basic result from martingale theory and a useful version of Fubini's theorem.

A.1 A filtered probability space

In the mathematical theory of probability, one begins with an abstract space Ω, a σ-algebra $\mathcal{F}$ on Ω, and a probability measure P on $(\Omega, \mathcal{F})$. The pair $(\Omega, \mathcal{F})$ is called a *measurable space* and the triple $(\Omega, \mathcal{F}, P)$ is called a *probability space*. Individual points $\omega \in \Omega$ represent possible *outcomes* for some experiment (broadly defined) in which we are interested. Identifying an appropriate outcome space Ω is always the first step in probabilistic modeling. Then $\mathcal{F}$ specifies the set of all *events* (subsets of Ω) to which we are prepared to assign probability numbers. Finally, the probability numbers $P(\cdot)$ reflect the relative likelihood of various events, whatever that may be interpreted to mean, and their specification is the second major step in probabilistic modeling. In economics one frequently interprets the probability measure P as a quantification of the subjective uncertainty experienced by some rational economic agent. Most physical scientists feel the need for a stronger, objective interpretation related to physical frequency.

In this book we usually take as primitive a probability space $(\Omega, \mathcal{F}, P)$ and a family $\mathbb{F} = \{\mathcal{F}_t, \ t \geq 0\}$ of σ-algebras on Ω such that (a) $\mathcal{F}_t \subseteq \mathcal{F}$ for all $t \geq 0$ and (b) $\mathcal{F}_s \subseteq \mathcal{F}_t$ if $s \leq t$. It is usual to express (a) and (b) by saying that $\mathbb{F}$ is an increasing family of sub-σ-algebras, or a *filtration* of $(\Omega, \mathcal{F})$. As a model element, $\mathbb{F}$ shows how information arrives (how uncertainty is resolved) as time passes. One interprets $\mathcal{F}_t$ as the set of all events whose occurrence or nonoccurrence can be determined at time t. The four-tuple $(\Omega, \mathcal{F}, \mathbb{F}, P)$ is called a *filtered probability space*.

171

A probability space $(\Omega, \mathcal{F}, P)$ is said to be *complete* if, whenever $A \in \mathcal{F}$, $P(A) = 0$, and $B \subseteq A$, one has $B \in \mathcal{F}$ as well. (Of course, it must be that $P(B) = 0$ for all such B.) That is, $(\Omega, \mathcal{F}, P)$ is said to be complete if all subsets of P-null sets in $\mathcal{F}$ are also elements of $\mathcal{F}$.

A filtration $\mathbb{F} = \{\mathcal{F}_t, \ t \geq 0\}$ of a complete probability space $(\Omega, \mathcal{F}, P)$ is itself said to be complete if $\mathcal{F}_0$ contains all the P-null sets in $\mathcal{F}$. (Note that each $\mathcal{F}_t$, $t > 0$, must also contain all P-null sets in $\mathcal{F}$, because $\mathcal{F}_0 \subseteq \mathcal{F}_t$ for $t > 0$.) The filtration $\mathbb{F}$ is said to be right-continuous if

$$(\text{A.1}) \qquad \mathcal{F}_t = \bigcap_{s > t} \mathcal{F}_s \qquad \text{for all } t > 0.$$

Loosely speaking, (A.1) means that any information known immediately after t is also known at t. A filtered probability space $(\Omega, \mathcal{F}, \mathbb{F}, P)$ is said to satisfy *the usual conditions* (this vaguely humorous term is actually standard in stochastic process theory) if (i) the underlying probability space $(\Omega, \mathcal{F}, P)$ is complete, and (ii) the filtration $\mathbb{F}$ is both complete and right-continuous.

Except as noted immediately below, *all probability spaces encountered in this book are assumed to be complete, and all filtered probability spaces are assumed to satisfy the usual conditions.* The exceptional cases are those where a concrete probability space is specified, as in Theorem 1.1, and no further assumptions are needed for the limited results developed in that context.

The usual conditions are harmless technical assumptions that are often needed for strict mathematical correctness; in particular, a number of fundamental results that are cited without proof in this book require the usual conditions. Starting with any filtered probability space $(\Omega, \mathcal{F}, \mathbb{F}, P)$, one can perform the following substitutions to ensure that the usual conditions are satisfied. First, let

$$\mathcal{N} := \{A \subset \Omega : A \subseteq B \text{ for some } B \in \mathcal{F} \text{ with } P(B) = 0\},$$

and then replace $\mathcal{F}$ by the smallest σ-algebra containing $\mathcal{F} \cup \mathcal{N}$, denoted $\sigma(\mathcal{F} \cup \mathcal{N})$ and called the *P-completion of $\mathcal{F}$*. Second, for each $t \geq 0$ replace $\mathcal{F}_t$ by

$$\bigcap_{s > t} \sigma(\mathcal{F}_s \cup \mathcal{N}).$$

These substitutions do not change anything materially from a modeling standpoint, but they produce a new filtered probability space, called *the usual P-augmentation of* $(\Omega, \mathcal{F}, \mathbb{F}, P)$, that satisfies the usual conditions.

Let $(\Omega, \mathcal{F}, \mathbb{F}, P)$ be a filtered probability space. A *stopping time* is a measurable function T from $(\Omega, \mathcal{F})$ to $[0, \infty]$ such that $\{\omega \in \Omega : T(\omega) \leq t\} \in \mathcal{F}_t$ for all $t \geq 0$. This definition involves the filtration in a fundamental way: one should really say that T is a stopping time *with respect to* $\mathbb{F}$. It is often useful to think of T as a plan of action; our definition requires that the decision to stop at or before time t depends only on information available at t. Now let $\mathcal{F}_T$ consist of all events $A \in \mathcal{F}$ such that

(A.2) $$\{\omega \in \Omega : \omega \in A \text{ and } T(\omega) \leq t\} \in \mathcal{F}_t$$

for all $t \geq 0$. Condition (A.2) is more compactly expressed by saying that $A \cap \{T \leq t\} \in \mathcal{F}_t$, and this level of symbolism will be used hereafter. One may think of $\mathcal{F}_T$ as the set of all events whose occurrence or nonoccurrence is known at the time of stopping.

A.2 Random variables and stochastic processes

Recall that $\mathbb{R}$ denotes the real line and $\mathcal{B}$ is the Borel σ-algebra on $\mathbb{R}$ (the smallest σ-algebra containing all the open sets). Given a probability space $(\Omega, \mathcal{F}, P)$, a *random variable* is a measurable function X from $(\Omega, \mathcal{F})$ to $(\mathbb{R}, \mathcal{B})$. Thus to each outcome $\omega \in \Omega$ there corresponds a numerical value $X(\omega)$, which we call the *realization* of X for outcome ω. One may identify or define many different random variables on a single outcome space, this identification reflecting different aspects of the experimental outcome that are of interest to the model builder. The *distribution* of X is defined as the probability measure Q on $(\mathbb{R}, \mathcal{B})$ given by

(A.3) $$Q(A) := P\{\omega \in \Omega : X(\omega) \in A\}, \qquad A \in \mathcal{B}.$$

The corresponding *distribution function* F is given by

$$F(x) := P\{\omega \in \Omega : X(\omega) \leq x\} := Q(-\infty, x], \qquad x \in \mathbb{R}.$$

It is known that F uniquely determines Q. Because the notion of function is more elementary than that of measure, it is usual to speak in terms of F rather than Q, but it will be seen shortly that the definition of the latter generalizes more readily.

It is customary to define a (one-dimensional) stochastic process as a family of random variables $X = \{X_t, t \in \mathcal{T}\}$, where $\mathcal{T}$ is an arbitrary index set. (Elements of $\mathcal{T}$ usually represent different points in time.) For our purposes, this definition will be specialized in three ways. First, the index set (or time domain) will always be $\mathcal{T} = [0, \infty)$. Second, all the processes considered in this book are right-continuous, that is, $X(\cdot, \omega) : [0, \infty) \to \mathbb{R}$ is right-continuous for almost all ω. Finally, attention will be restricted to processes X that are *jointly measurable*, which means that $X : \Omega \times [0, \infty) \to \mathbb{R}$ is measurable with respect to the product σ-algebra $\mathcal{F} \times \mathcal{B}[0, \infty)$, where $\mathcal{B}[0, \infty)$ is the Borel σ-algebra on $[0, \infty)$. To denote the value of X at a point $(\omega, t) \in \Omega \times [0, \infty)$ we write either $X(\omega, t)$ or $X_t(\omega)$, with a consistent preference for the latter notation. It is a standard result in measure theory (usually stated as part of Fubini's theorem) that $X(\omega) := \{X_t(\omega), t \geq 0\}$ is a Borel-measurable function $[0, \infty) \to \mathbb{R}$ for each fixed ω. Similarly, $X_t : \Omega \to \mathbb{R}$ is an $\mathcal{F}$-measurable function (a random variable) for each fixed t. The function $X(\omega)$ is called the realization, or trajectory, or *sample path* of the process X corresponding to outcome ω.

To repeat, the term "stochastic process" (often shortened to just "process"), as it is used in this book, automatically implies right-continuity and joint measurability. A number of standard results that are stated in the book require one or both of those standing assumptions for their validity, but the standing assumption(s) will not be repeated when such results are stated.

Our next topic is continuous processes, for which some preliminary definitions are necessary. Let $C := C[0, \infty)$ be the space of all continuous functions $x : [0, \infty) \to \mathbb{R}$. (Functions are here denoted by letters like x and y, rather than the usual f and g because we are thinking of them as sample paths of stochastic processes.) The standard metric ρ on this space is defined by

$$
(A.4) \qquad \rho(x, y) := \sum_{n=1}^{\infty} \frac{\left(\frac{1}{2}\right)^n \rho_n(x, y)}{1 + \rho_n(x, y)},
$$

where

$$
(A.5) \qquad \rho_t(x, y) := \sup_{0 \leq s \leq t} |x(s) - y(s)| \qquad t \geq 0
$$

for $x, y \in C$. Note that ρ_t is the usual metric of uniform convergence on $C[0, t]$. When we say that $x_n \to x$ in C, this means that $\rho(x_n, x) \to 0$ as $n \to \infty$. The following is immediate from (A.5) and (A.4).

Proposition A.1 $x_n \to x$ in C if and only if $\rho_t(x_n, x) \to 0$ as $n \to \infty$ for all $t > 0$.

One may paraphrase Proposition A.1 by saying that ρ induces on C the topology of uniform convergence on finite intervals. A subset A of C is said to be *open* if for every point $x \in A$ there exists a radius $r > 0$ such that all $y \in C$ with $\rho(x, y) < r$ belong to A. As a precise analog of $\mathcal{B}$, we define $\mathcal{C}$

as the smallest σ-algebra on C containing all the open sets, calling $\mathcal{C}$ the Borel σ-algebra on C.

A stochastic process X on $(\Omega, \mathcal{F}, P)$ is said to be *continuous* if $X(\omega) \in C$ for almost all $\omega \in \Omega$. From this and the fact that X_t is measurable with respect to $\mathcal{F}$ for each $t \geq 0$, it can be shown that X is a measurable mapping $(\Omega, \mathcal{F}) \to (C, \mathcal{C})$. That is, a continuous process may be viewed as a *random element* of the metric space C. The *distribution* of a continuous process X is the probability measure Q on $(C, \mathcal{C})$ defined by (A.3) with $\mathcal{C}$ in place of $\mathcal{B}$. One may paraphrase this definition by calling Q the probability measure on $(C, \mathcal{C})$ *induced from P by X*. It may be desirable to elaborate on this critically important concept. It can be verified that the sets

$$A := \{x \in C : x(T) \leq a\}$$

and

$$B := \{x \in C : x(t) \leq b, \ 0 \leq t \leq T\}$$

are both elements of $\mathcal{C}$. (Here, a, b and $T > 0$ are constants.) Applying the definition of Q, we have

$$Q(A) = P\{\omega \in \Omega : X_T(\omega) \leq a\}$$

and

$$Q(B) = P\{\omega \in \Omega : M_T(\omega) \leq b\}$$

where $M_T(\omega) := \sup\{X_t(\omega), 0 \leq t \leq T\}$. Suppressing the dependence on ω, as is usual in probability theory, these relations can be written as

$$Q(A) = P\{X_T \leq a\} \quad \text{and} \quad Q(B) = P\{M_T \leq b\}.$$

Thus knowledge of the process distribution Q gives us, at least in principle, not only the distributions of the individual random variables X_T but also those of more complex functionals like maxima. It is an important fact that the process distribution Q is uniquely determined by the *finite-dimensional distributions* of X. This result will not be used here but interested readers are referred to Chapter 2 of Billingsley (1999) for further discussion.

Throughout this section we have spoken of a stochastic process X defined on some probability space $(\Omega, \mathcal{F}, P)$. Enriching this setting, suppose that the probability space is equipped with a filtration $\mathbb{F} = \{\mathcal{F}_t, \ t \geq 0\}$. We say that X is an *adapted* process on the filtered probability space $(\Omega, \mathcal{F}, \mathbb{F}, P)$ if X_t is measurable with respect to $\mathcal{F}_t$ for all $t \geq 0$. Heuristically, this means that the information available at time t includes the history of X up to that point.

Let X be a stochastic process defined on some (complete) probability space $(\Omega, \mathcal{F}, P)$, and for each $t \geq 0$ let $\mathcal{G}_t$ be the smallest sub-σ-algebra of $\mathcal{F}$ with respect to which all the random variables $\{X_s, 0 \leq s \leq t\}$ are measurable. We denote by $\mathbb{F} = \{\mathcal{F}_t, t \geq 0\}$ the usual P-augmentation of $\{\mathcal{G}_t, t \geq 0\}$, calling this the *filtration generated by* X. The filtered probability space $(\Omega, \mathcal{F}, \mathbb{F}, P)$ automatically satisfies the usual conditions.

A.3 A canonical example

To give a concrete example of a filtered probability space and a stochastic process on that space, we take $\Omega = C$ and define the *projection map* X_t : $C \to \mathbb{R}$ via

$$X_t(\omega) = \omega(t) \qquad \text{for } t \geq 0 \text{ and } \omega \in C.$$

(Here the generic element of C is denoted by a lower case Greek ω rather than the lower case Roman x used earlier, for obvious reasons.) Let P be an arbitrary probability measure on $(C, \mathcal{C})$, let $\mathcal{F}$ be the P-completion of $\mathcal{C}$, and let $\mathbb{F} = \{\mathcal{F}_t, t \geq 0\}$ be the filtration generated by the process X. Then $(\Omega, \mathcal{F}, \mathbb{F}, P)$ is a filtered probability space satisfying the usual conditions, and X is an adapted continuous process on that space.

This canonical setup is appropriate when the sample path of X is the only relevant source of uncertainty for current purposes, and the only relevant information available at time t is the history of X up to that point. (In general, when we say that a process X is adapted to a filtration $\mathbb{F}$, the σ-algebra $\mathcal{F}_t$ may contain more information than just the history of X up to time t.) Note that the coordinate process X, viewed as a mapping $C \to C$, is the identity map $X(\omega) = \omega$.

A.4 Two martingale theorems

Let X be a stochastic process on some filtered probability space $(\Omega, \mathcal{F}, \mathbb{F}, P)$. This process X is said to be a *martingale* if it is adapted, $E(|X_t|) < \infty$ for all $t \geq 0$ and $E(X_t | \mathcal{F}_s) = X_s$ whenever $s \leq t$. This is another definition that involves the filtration in a fundamental way.

Let X be a process defined on some probability space $(\Omega, \mathcal{F}, P)$. When we say that X is a martingale *with respect to its own filtration*, or say that X is a martingale without specifying a filtration, that means the following:

X is a martingale on the filtered probability space $(\Omega, \mathcal{F}, \mathbb{F}, P)$, where $\mathbb{F} = \{\mathcal{F}_t, \ t \geq 0\}$ is the filtration generated by X itself (see Section A.2).

From the vast theory of martingales we will need just two basic results. Theorem A.2 below is a special case of the martingale convergence theorem appearing on page 16 of Chung and Williams (1990), and Theorem A.3 is a special case of Doob's optional sampling theorem (Karatzas and Shreve, 1998, p. 19).

Theorem A.2 (Martingale Convergence Theorem) *If X is a martingale on some filtered probability space $(\Omega, \mathcal{F}, \mathbb{F}, P)$ and if $\sup\{E(X_t^2), \ t \geq 0\} < \infty$, then $X_\infty := \lim X_t$ exists almost surely and $X_t \to X_\infty$ in the L^2 sense.*

Theorem A.3 (Martingale Stopping Theorem) *Let X be a martingale and T a stopping time on some filtered probability space $(\Omega, \mathcal{F}, \mathbb{F}, P)$. Then the stopped process $\{X(t \wedge T), \ t \geq 0\}$ is also a martingale.*

From Theorem A.3 is it immediate that $E[X(t \wedge T)] = E[X(0)]$ for any $t > 0$. If $P\{T < \infty\} = 1$ (hereafter written simply $T < \infty$), then of course $(t \wedge T) \to T$ almost surely as $t \to \infty$. However, it is *not* necessarily true that $E[X(t \wedge T)] \to E[X(T)]$ as $t \to \infty$, and hence not necessarily true that $E[X(T)] = E[X(0)]$. For example, let X be a standard Brownian motion (zero drift and unit variance) with $X(0) = 0$, and let T be the first time t at which $X(t) = 1$. It is well known that X is a martingale, that T is a stopping time, and that $T < \infty$, but it is obviously false that $E[X(T)] = E[X(0)]$. The following proposition gives an easy sufficient condition for $E[X(T)] = E[X(0)]$.

Corollary A.4 *In addition to the hypotheses of Theorem A.3, suppose that $T < \infty$ and the stopped process $\{X(t \wedge T), \ t \geq 0\}$ is uniformly bounded. Then $E[X(T)] = E[X(0)]$.*

Proof Because $T < \infty$ almost surely, $X(t \wedge T) \to X(T)$ almost surely as $t \to \infty$. Because $\{X(t \wedge T), \ t \geq 0\}$ is bounded by hypothesis, the bounded convergence theorem gives $E[X(T \wedge t)] \to E[X(T)]$ as $t \to \infty$. But from Proposition A.1 we have $E[X(t \wedge T)] = E[X(0)]$ for all $t \geq 0$ and hence $E[X(T)] = E[X(0)]$. $\qquad\square$

A.5 A version of Fubini's theorem

Recall the general statement of Fubini's theorem presented in Billingsley (1995) and other basic texts on measure theory. Let $\mathcal{B}[0, \infty)$ be as in Section A.2, let λ be Lebesgue measure, and let X be a stochastic process on

$(\Omega, \mathcal{F}, P)$. Specializing Fubini's theorem to the product space $\Omega \times [0, \infty)$, the product measure $P \times \lambda$, and the jointly measurable function $X : \Omega \times [0, \infty) \to \mathbb{R}$, we get the following theorem.

Theorem A.5 *If $E[\int_0^\infty |X(t)|\, dt] < \infty$, then*

$$(A.6) \qquad\qquad E\left[\int_0^\infty X(t)\, dt\right] = \int_0^\infty E\left[X(t)\right]\, dt.$$

A closely related result, sometimes called *Tonelli's theorem*, says that (A.6) holds for positive processes X without any further hypotheses; in particular the iterated integrals on the two sides are either both infinite or else both finite and equal.

Appendix B

Real Analysis

This appendix collects several results from real analysis that play a central role in the text. All of the cited results can be found in Bartle (1976) or Billingsley (1995), or else are easy generalizations of results proved in those books.

B.1 Absolutely continuous functions

Let $f : [0, \infty) \to \mathbb{R}$ be fixed. The function f is said to be absolutely continuous on $[0, t]$ if, given $\epsilon > 0$, there is a $\delta > 0$ such that

$$\sum_{i=1}^{n} |f(b_i) - f(a_i)| < \epsilon$$

for every finite collection of nonoverlapping intervals $\{(a_i, b_i); i = 1, \ldots, n\}$ with $0 \leq a_i \leq b_i \leq t$ and

$$\sum_{i=1}^{n} (b_i - a_i) < \delta.$$

When we say that f is absolutely continuous, this means that it is absolutely continuous on $[0, t]$ for every $t > 0$.

Proposition B.1 *f is absolutely continuous if and only if there is a measurable function $g : [0, \infty) \to \mathbb{R}$ such that $f(t) = f(0) + \int_0^t g(s) \, ds$ (Lebesgue integral), $t \geq 0$.*

The function g appearing in Proposition B.1 is called a *density* for f; it is not unique, but any two densities must be equal except on a set of Lebesgue measure zero. An absolutely continuous function is differentiable almost everywhere (Lebesgue measure) and the derivative is a density.

B.2 VF functions

Again let $f : [0, \infty) \to \mathbb{R}$ be fixed. The total variation of f over $[0, t]$ is defined as

$$v_t(f) := \sup \left\{ \sum_{i=1}^{n} |f(t_i) - f(t_{i-1})| \right\}$$

where the supremum is taken over all finite partitions $0 = t_0 < \cdots < t_n = t$. We call f a VF function if $v_t(f) < \infty$ for all $t > 0$. (The acronym VF comes from the French literature on stochastic processes where it stands for *variation finite*.)

Proposition B.2 *g is a VF function on $[0, \infty)$ if and only if it can be written as the difference of two increasing functions on $[0, \infty)$.*

B.3 Riemann–Stieltjes integration

Starting with two functions $f, g : [0, \infty) \to \mathbb{R}$, recall from Section 29 of Bartle (1976) what it means for f to be integrable with respect to g over $[0, t]$. This is analogous to the more familiar definition of Riemann integrability; $\int f \, dg$ will not exist unless f and g have a certain amount of structure. The following results can be found in Sections 29 and 30, respectively, of Bartle (1976).

Theorem B.3 (Integration by Parts Theorem) *f is integrable with respect to g over $[0, t]$ if and only if g is integrable with respect to f over $[0, t]$. In this case,*

$$\text{(B.1)} \qquad \int_0^t f \, dg = [f(t)g(t) - f(0)g(0)] - \int_0^t g \, df.$$

Theorem B.4 (Integrability Theorem) *If f is continuous and g is increasing over $[0, t]$ then f is integrable with respect to g over $[0, t]$.*

In (B.1) and hereafter, we write $\int_0^t$ to signify a Riemann–Stieltjes integral over $[0, t]$. In the integral on the left side of (B.1), we call f the *integrand* and g the *integrator*. The indefinite integral $\int f \, dg$ will be understood to signify a function $h : [0, \infty) \to \mathbb{R}$ defined by

$$h(t) := \int_0^t f \, dg \qquad \text{for } t \geq 0.$$

When we say that $\int f \, dg$ exists, this means that f is integrable with respect to g over every finite interval $[0, t]$. By combining Proposition B.2, Theorem B.3, and Theorem B.4, we arrive at the following important result.

Proposition B.5 *If f is continuous and g is a VF function, then $\int f \, dg$ and $\int g \, df$ both exist, and the integration by parts formula (B.1) is valid for all $t \geq 0$.*

B.4 The Riemann–Stieltjes chain rule

The following result does not appear in Bartle (1976), but one can easily construct a proof by generalizing that of Bartle's Theorem 30.13.

Proposition B.6 (Chain Rule) *Suppose that $f, g : [0, \infty) \to \mathbb{R}$ are continuous, that g is a VF function, and that $\phi : \mathbb{R} \to \mathbb{R}$ is continuously differentiable. Then*

(B.2)
$$\int_0^t f \, d\phi(g) = \int_0^t f\phi'(g) \, dg, \qquad t \geq 0.$$

In formula (B.2) we write $\phi(g)$ to denote the function that has value $\phi(g(t))$ at time t. Similarly, $f\phi'(g)$ denotes the function that has value $f(t) \, \phi'(g(t))$ at time t; the right side of (B.2) is the integral of this function with respect to g. It is immediate from Proposition B.5 that the integrals on both sides of (B.2) exist. One may state (B.2) in more compact differential form as

(B.3)
$$d\phi(g) = \phi'(g) \, dg$$

with the understanding that this is just shorthand for (B.2). Because (B.3) generalizes the familiar chain rule for differentiating the composition of two functions, we shall hereafter refer to (B.2) as the *Riemann–Stieltjes chain rule*.

Let X be a continuous stochastic process on some probability space, and further suppose that $X(\omega)$ is a VF function for all $\omega \in \Omega$. (One may express this state of affairs more succinctly by saying that X is a *continuous VF process*.) For each fixed ω, apply (B.2) with $X(\omega)$ in place of g and $f(t) = 1$ for all t. The left side then reduces to $\phi(X(t)) - \phi(X(0))$ and we arrive at the sample path relationship

(B.4)
$$\phi(X(t)) = \phi(X(0)) + \int_0^t \phi'(X) \, dX, \qquad t \geq 0.$$

To repeat, (B.4) is a statement of equality between *random variables*; in the usual way, we suppress the dependence of X on ω to simplify typography. It is the purpose of Itô's formula, on which we focus in Chapter 4, to develop an analog of (B.4) for certain continuous processes X that do *not* have VF sample paths.

B.5 Notational conventions for integrals

Where we have written $\int f\,dg$ to denote the Riemann–Stieltjes integral of f with respect to g, some authors write $\int f(t)\,dg(t)$. Because the latter notation involves so many unnecessary symbols, we shall use it only to show special structure for f of g. For example, the expressions

(B.5) $$\int_0^T f\,dt$$

(B.6) $$\int_0^T e^{-\lambda t}\,dg$$

(B.7) $$\text{and}\quad \int_0^T e^{-\lambda t}h\,dt$$

may be written to signify integrals over $[0, T]$ where: in (B.5) the integrand is f and the integrator is $g(t) = t$; in (B.6) the integrand is $f(t) = \exp(-\lambda t)$; and in (B.7) the integrand is $f(t) = \exp(-\lambda t)h(t)$ and the integrator is $g(t) = t$. Occasionally, to conform with the usual notation for Riemann integrals, expressions (B.5) and (B.7) may be written as

$$\int_0^T f(t)\,dt \quad\text{and}\quad \int_0^T e^{-\lambda t}h(t)\,dt.$$

References

Bartle, Robert G. 1976. *The Elements of Real Analysis*. Second edn. John Wiley & Sons, New York-London-Sydney. (Cited on pages xi, 179, 180, and 181.)

Bather, John, and Chernoff, Herman. 1967. Sequential Decisions in the Control of a Space-Ship (Finite Fuel). *J. Appl. Probab.*, **4**(3), 584–604. (Cited on page 135.)

Beneš, V. E., Shepp, L. A., and Witsenhausen, H. S. 1980. Some solvable stochastic control problems. *Stochastics*, **4**, 39–83. (Cited on page 135.)

Bensoussan, A., and Lions, J. L. 1975. Nouvelles Methodes en Contrôle Impulsionnel. *Applied Mathematics and Optimization*, **1**, 289–312. (Cited on page 134.)

Billingsley, Patrick. 1995. *Probability and Measure*. Third edn. Wiley Series in Probability and Mathematical Statistics. New York: Wiley-Interscience. (Cited on pages xi, 5, 177, and 179.)

Billingsley, Patrick. 1999. *Convergence of Probability Measures*. Second edn. Wiley Series in Probability and Statistics. New York: Wiley-Interscience. (Cited on pages 2 and 175.)

Breiman, Leo. 1968. *Probability*. Reading, MA: Addison-Wesley Publishing Company. (Cited on page 2.)

Chernoff, Herman. 1968. Optimal stochastic control. *Sankhya A*, **30**(3), 221–252. (Cited on page 148.)

Chung, Kai Lai, and Williams, Ruth J. 1990. *Introduction to Stochastic Integration*. Second edn. Probability and its Applications. Boston: Birkhaäuser. (Cited on pages 4, 7, 8, 67, and 177.)

Çinlar, Erhan. 1975. *Introduction to Stochastic Processes*. Englewood Cliffs, NJ: Prentice-Hall Inc. (Cited on pages 96 and 97.)

Curtiss, J. H. 1942. A note on the theory of moment generating functions. *Ann. Math. Statist.*, **13**(4), 430–433. (Cited on page 7.)

Daley, Brendan, and Green, Brett. 2012. Waiting for news in the market for lemons. *Econometrica*, **80**(4), 1433–1504. (Cited on page 161.)

Dayanik, Savaş, and Karatzas, Ioannis. 2003. On the optimal stopping problem for one-dimensional diffusions. *Stoch. Proc. Appl.*, **107**, 173–212. (Cited on pages 85, 86, 88, and 134.)

DeGroot, Morris H. 2004. *Optimal Statistical Decisions*. Wiley Classics Library. Hoboken, NJ: Wiley-Interscience. (Cited on page 148.)

Dixit, Avinash. 1993. *The Art of Smooth Pasting*. Chur, Switzerland: Harwood Academic Publishers. (Cited on page x.)

Dixit, Avinash K., and Pindyck, Robert S. 1994. *Investment Under Uncertainty*. Princeton, NJ: Princeton University Press. (Cited on page 88.)

Egami, Masahiko. 2008. A direct solution method for stochastic impulse control problems of one-dimensional diffusions. *SIAM J. Control. Optimizat.*, **47**(3), 1191–1218. (Cited on page 134.)

El Karoui, N., and Chaleyat-Maurel, M. 1978. Un problème de réflexion et ses aplications au temps local et aux équations différentielles stochastiques sur ℝ–Cas continu. *Astérisque*, **52-53**, 117–144. (Cited on page 20.)

Emanuel, David C., Harrison, J. Michael, and Taylor, Allison J. 1975. A diffusion approximation for the ruin function of a risk process with compounding assets. *Scand. Actuar. J.*, **1975**(4), 240–247. (Cited on pages 157 and 158.)

Harrison, J. Michael. 1973. Assembly-like queues. *J. Appl. Probab.*, **10**, 354–367. (Cited on page 34.)

Harrison, J. Michael. 1977. Ruin problems with compounding assets. *Stoch. Proc. Appl.*, **5**(1), 67–79. (Cited on page 158.)

Harrison, J. Michael. 1978. The diffusion approximation for tandem queues in heavy traffic. *Adv. Appl. Probab*, **10**, 886–905. (Cited on page 32.)

Harrison, J. Michael, and Lemoine, Austin J. 1981. Sticky Brownian motion as the limit of storage processes. *J. Appl. Probab.*, **18**(1), 216–226. (Cited on page 165.)

Harrison, J. Michael, and Reiman, Martin I. 1981. Reflected Brownian motion on an orthant. *Ann. Probab.*, **9**(2), 302–308. (Cited on page 108.)

Harrison, J. Michael, and Taksar, Michael I. 1983. Instantaneous control of Brownian motion. *Math. Oper. Res.*, **8**(3), 439–453. (Cited on pages 110 and 134.)

Harrison, J. Michael, and Taylor, Allison J. 1978. Optimal control of a Brownian storage system. *Stoch. Proc. Appl.*, **6**(2), 179–194. (Cited on page 134.)

Harrison, J. Michael, Sellke, Thomas M., and Taylor, Allison J. 1983. Impulse control of Brownian motion. *Math. Oper. Res.*, **8**(3), 454–466. (Cited on page 134.)

Karatzas, Ioannis, and Shreve, Steven E. 1991. *Methods of Mathematical Finance*. New York: Springer. (Cited on page 88.)

Karatzas, Ioannis, and Shreve, Steven E. 1998. *Brownian Motion and Stochastic Calculus*. 2nd edn. New York: Springer. (Cited on pages 12, 52, 60, and 177.)

Kruk, Lukasz, Lehoczky, John, Ramanan, Kavita, and Shreve, Steven. 2007. An explicit formula for the Skorokhod map on $[0, a]$. *Ann. Probab.*, **35**(5), 1740–1768. (Cited on page 22.)

Liptser, Robert S., and Shiryayev, Albert N. 2000a. *Statistics of Random Processes I*. 2nd rev. and exp. edn. Springer. Translated by A. B. Aries. (Cited on page 150.)

Liptser, Robert S., and Shiryayev, Albert N. 2000b. *Statistics of Random Processes II*. 2nd rev. and exp. edn. Springer. Translated by A. B. Aries. (Cited on page 148.)

McDonald, Robert, and Siegel, Daniel. 1986. The value of waiting to invest. *Quart. J. Econ.*, **101**(4), 707–728. (Cited on pages x, 77, and 88.)

McKean, Jr., Henry P. 1969. *Stochastic Integrals*. New York: Academic Press. (Cited on page 53.)

Øksendal, Bernt K. 2007. *Stochastic Differential Equations: An Introduction with Applications*. 6 edn. Berlin; New York: Springer. (Cited on pages 60 and 88.)

Ormeci, Melda, Dai, J. G., and Vande Vate, John H. 2008. Impulse control of Brownian motion: The constrained average cost case. *Operations Research*, **56**, 618–629. (Cited on pages 126 and 134.)

Peskir, Goran, and Shiryaev, Albert N. 2006. *Optimal Stopping and Free-Boundary Problems*. Lectures in Mathematics. ETH Zürich. Birkhäuser Basel. (Cited on pages 88 and 154.)

Poor, H. Vincent, and Hadjiliadis, Olympia. 2008. *Quickest Detection*. Cambridge: Cambridge University Press. (Cited on page 8.)

Rogers, L. C. G., and Williams, David. 1987. *Diffusions, Markov Processes and Martingales: Itô Calculus*. Wiley Series in Probability and Mathematics Statistics, vol. 2. New York: John Wiley & Sons Ltd. (Cited on page 8.)

Ross, S.M. 1983. *Stochastic Processes*. New York: Wiley. (Cited on pages 98 and 104.)

Stokey, Nancy L. 2009. *The Economics of Inaction*. New Jersey: Princeton University Press. (Cited on page x.)

Wenocur, M. 1982. *A production network model and its diffusion limit*. Ph.D. thesis, Department of Statistics, Stanford University. (Cited on page 34.)

Wonham, W. 1964. Some applications of stochastic differential equations to optimal nonlinear filtering. *SIAM J. Contr. Optimizat.*, **2**(3), 347–369. (Cited on page 150.)

Index

187

Printed in the United States
By Bookmasters